8

The Neuroscientific Turn

Transdisciplinarity in the Age of the Brain

Melissa M. Littlefield
and
Jenell M. Johnson
Editors

The University of Michigan Press
Ann Arbor

Published in the United States of America by
The University of Michigan Press
Manufactured in the United States of America
⊚ Printed on acid-free paper

2015   2014   2013   2012   4   3   2   1

A CIP catalog record for this book is available from the British Library.

Library of Congress Cataloging-in-Publication Data

The neuroscientific turn : transdisciplinarity in the age of the brain / Melissa M. Littlefield
    and Jenell M. Johnson, editors.
        p. ; cm.
    Includes bibliographical references and index.
        ISBN 978-0-472-11826-7 (hardback : alk. paper) — ISBN 978-0-472-02835-1
    (e-book)
        I. Littlefield, Melissa M., 1979– II. Johnson, Jenell M., 1978–
    [DNLM: 1. Neuroscience. WL 100]
    612.8—dc23   2012011048

To Susan M. Squier, for opening doors

Acknowledgments

We are, first and foremost, thankful to our collaborators in this project: the authors who wrote and revised essays for this volume, and who displayed their willingness to transgress boundaries, to be innovative, and to ask important questions. Our sincere gratitude to those who read part or all of the manuscript: Spencer Schaffner, Justine Murison, Kate Viera, and the anonymous reviewers. We would also like to thank Andrew Shail, Bob Markley, Susan Squier, Hugh Crawford, and Laura Otis for offering advice along the way. Thanks also to Martyn Pickersgill and Ira van Keulen for helping to shape our ideas about popular neuroscience. Finally, thanks to our terrific research assitants, Jessica Mercado and Megan Condis.

Beyond these individuals, we are indebted to two organizations: the Society for Literature, Science, and the Arts (SLSA) and the European Neuroscience and Society Network (ENSN). For the past decade, SLSA has provided both the occasion to delve into many of the questions that inform this collection and the audience to engage with them. SLSA was literally our meeting place: where we pored over submitted abstracts among coffee and crumbs in Atlanta, where we presented our ideas for the collection on a panel with Sarah Birge, where we pitched the book to Tom Dwyer over lunch in Indianapolis. ENSN has been our source of all things "neuro," from Neuroschool to fMRI experiments, and the people we have met through this collaborative effort have forever changed our perspectives about what it means to work interdisciplinarily—or better yet, transdisciplinarily. In particular, we would like to thank Nikolas Rose, Giovanni Frazzetto, Klaus-Peter Lesch, and Andreas Roepsdorff for their insight, for opening their laboratories, and for loosening their purse strings so that we could learn about neuroscience and work with technologies like MRI scanners and PCR.

In addition, we want to offer our thanks to Tom Dwyer, Alexa Ducsay,

Christina Milton, and members of the University of Michigan Press's editorial and production staff for producing a beautiful volume. This collection would not be possible without the support of the University of Illinois Research Board and the Department of English. Finally, we would also like to acknowledge and commemorate Susan Leigh Star, who passed away in March 2010. Leigh had agreed to contribute an essay to this volume, and, though the collection is full of wonderful chapters, hers is sorely missed.

Melissa would like to thank Jenell for her tireless work, her amazing scholarship, and her sense of humor. This volume was the collaboration of friends long before it was a scholarly collection. Thanks also to Des Fitzgerald, Martin Dietz, Kasper Knudsen, and James Tonks for collaborating in all things fMRI; to the graduate students in my Neuroscience and Humanities course for their insightful comments; to John Littlefield, Valerie Perialas, and Jonathan Littlefield for their tireless support; and, finally, to Isaac Sosnoff and Spencer Schaffner: Σ'αγαπώ.

Jenell would like to first thank Melissa: trusted coauthor, meticulous coeditor, and dear friend. Thanks also to Alison Cool, Daniel Stjepanović, Morten Bülow, and my other Würzburg neuroschoolmates; my colleagues at Louisiana State University (especially Lillian Bridwell-Bowles) and the University of Wisconsin-Madison (especially Rob Asen, Karma Chávez, Jonathan Gray, Rob Howard, Eunjung Kim, Steve Lucas, Sara McKinnon, Ellen Samuels, Walt Schalick, and Sue Zaeske). My gratitude to Megan Zuelsdorff for countless hours of brain-y conversations and for introducing me to neurotheology many years ago. Thanks to my family: James Johnson, Joy Johnson, Jenna Johnson, Jodi Carreon, and Ryan Carreon, for their constant support. And, finally, my deepest thanks to Michael Xenos, my best critic and biggest fan.

Contents

Preface

A Neuro-Pivot

Judy Illes

In an essay that I wrote for the *American Journal of Bioethics-Neuroscience* when I stepped down as editor in 2009, I reflected with enthusiasm on the rapid growth of the discipline of neuroethics. I also took a few punches at the rate and breadth of that growth, and neuro-terms that had surfaced, at least in part, as a result. Some of the serious and straightforward ones I highlighted were *neuroeducation, neurolaw, neuroeconomics, neuroepistemology,* and *neuromarketing.* These have taken root in the peer-reviewed literature and have even become associated with their own professional and remarkably robust domains. More lighthearted and perhaps provocative were *neurosurrealism, neurofantasy, neuromyths, neurotime,* and *neuroage* that variously capture the movement of brain science into our lives—sometimes with hope and often with hype—and on a daily basis at that. My own neologism was *neurologism,* borrowing from the term used to describe the breakdown of language in patients with temporal lobe pathology, such as those with Wernicke's jargonaphasia. These patients produce largely unintelligible sentences with words and phonemes that resemble the person's language of origin, assembled in seemingly unaffected syntax. Here, in Melissa M. Littlefield and Jenell M. Johnson's "Introduction: Theorizing the Neuroscientific Turn—Critical Perspectives on a Translational Discipline," we discover yet another neologistic neurologism: Neuro-Turn, short for Neuroscientific Turn. What could this neologism represent? Unlike the neurologic patients with whom we are sometimes never able to recover meaningful communication or engagement through language, Neuro-turn is perhaps the most clear and all-encompassing neologism yet; the most powerful new kid on the neuroethics block. Allow me to explain why.

Neuro-Turn, the word, and *Neuroscientific Turn*, the volume, are connected by three premises: (1) neuroscience has a long and dynamic history that must be understood contextually; (2) neuroscience has become a translational discipline; and, as such, (3) neuroscience is transformative. They reflect the revolution we are experiencing in how we study the brain and in how we live with what it gives us. The flow of information has become multidimensional, multidirectional, and, if I dare say so, multi-meaningful. *Neuro-turn* manages to capture not only the complexities of integrating new knowledge about the brain into the humanities, society, and technology but the many different directions that integration can take.

Some of these directions are pragmatic. They urge us to critically interrogate different meanings and identities, and to be wary of hype (however beautiful a brain image might be). Other directions are more theoretical, reminding us about the inextricable relationships between thoughts and styles, between religions and cultures, and that what we can do we need not necessarily ought to. These directions therefore mirror a movement that is both academic and social, on the one hand, and combining and colliding, on the other.

Like in the free market, competition is good. Collisions, in fact, keep us at least as sharp as our collaborations, if not more so. At a meeting of the European Molecular Biology Organization in Heidelburg, Germany, in 2006, Professor Jean Pierre Changeux challenged me about the need for such movement and the value added to new terms such as *neuroethics* that we were giving it. My answer was simple: it hardly matters what we call it, as long as we do it (Illes 2007). Nomenclature and semantics pale as points of argument in comparison to the rigor with which research methodologies must be chosen and exploited, and hypotheses explored. Indeed, at a time when we are witness to one of the most reductionist calls in neuroscience ever—for exploration of the connectome—new directions will be shaped but they must not be hijacked by momentum. Concepts like autonomy will be transformed, lives improved, cultures understood.

This is *Neuro-Turn*'s time. As Murison writes, to examine a "'turn,' . . . is to give a particular shape to intellectual history. It is a 'profound pivot.'" This neuroscientific turn is that pivot for the field of neuroethics. It is a departure from well-trodden discussions and debates about neuroimaging for mind-reading, lifestyle drugs and cognitive enhancement, free will and responsibility. It is fresh, exciting, and inspiring.

WORKS CITED

Illes, J. "Neurologisms." *American Journal of Bioethics* 9, no. 9 (2009): 1.
Illes, J. "Empirical Neuroethics. Can Brain Imaging Visualize Human Thought? Why Is Neuroethics Interested in Such a Possibility?" *EMBO Reports* 8 (2007): S57–61.

Introduction

Theorizing the Neuroscientific Turn—Critical
Perspectives on a Translational Discipline

Melissa M. Littlefield and Jenell M. Johnson

> Neuroscience is the new philosophy, some say, and there is no doubt
> that brain research is the hottest topic a scientist can dabble in, and
> the most distinguished thing you can put on your calling card. (Frank
> 2009, 9)

The application of neuroscience to fields beyond medicine[1] has been characterized as revolutionary, akin to the industrial and information revolutions (Lynch 2009, 10) and evidence of the birth of a "neurosociety" in
which all domains of life and knowledge production are (or soon will be)
under the sign of the "neuro." Each day, it seems, the popular press reports
new neuroscientific findings with breathless wonder. Recent headlines
have claimed that neuroscience has the power to read our minds (Sample
2007), erase our memories (Carey 2009), predict our propensity for violence (Harrell 2010), and alter the neural fabric of our identities. More so
than any other product of the neurosciences, studies using functional magnetic resonance imaging (fMRI) have found a particularly strong purchase
in the public imagination—strong to the point where a recent advertise-
ment for a memory-boosting supplement in the SkyMall catalog claimed
it could make your brain "light up like a Christmas tree," complete with a
before-and-after brain scan that illustrated the product's efficacy (too bad
the picture was in black and white). Consider too, the recent proliferation
of popular neuroscience on bookstore shelves, and the staggering popularity of these books for audiences who might not know a neuron from a glial

cell. Books by Oliver Sacks, Antonio Damasio, V. S. Ramachandran, Joseph LeDoux, and Steven Pinker rapidly have become best sellers, propelling popular neuroscience (Johnson and Littlefield 2011) into its own genre and turning scientists into superstars.

Lest this phenomenon of the "neurorevolution" appear to be the fleeting invention of popular culture, one need only look to the emergence of the many "neuro-disciplines" (Vidal 2009) proliferating in the academy, including but not limited to neuroeconomics, neurohistory, neuroanthropology, neuroaesthetics, neuromarketing, neurosociology, neuropolitics, neuroethics, neurotheology, and even the neurohumanities.[2] Scholars working in these burgeoning fields of inquiry come from varying disciplinary backgrounds. Some have chosen the neurosciences as a methodology while others, to borrow the cliché, have had neuroscience thrust upon them.[3] While it is too early to say whether neuroscience has—or will—become "the new philosophy" as suggested by our epigraph,[4] the ubiquitous adoption of neuroscience by multiple fields portends a significant phase shift in interdisciplinary research that we believe can readily be understood as the neuroscientific turn.[5]

The omnipresence of neuroscience has already begun to change the landscape of academic disciplinarity across numerous countries, including—but not limited to—the United States, Canada, the United Kingdom, Germany, and Denmark. In large part, the widespread interest in and potential affiliations between disciplines and nations have been made possible by interdisciplinary centers and collectives.[6] Take, for example, the European Neuroscience and Society Network (ENSN), which was funded for five years (2007–12) by the European Science Foundation and housed at the London School of Economics BIOS Centre; the ENSN boasts over 120 affiliates from the United States, Canada, and Europe. By providing collaborative funding opportunities and interdisciplinary events (including "neuroschools"),[7] the ENSN has helped to bridge gaps between the neurosciences, the humanities, and the social sciences. Scholarship produced by ENSN affiliates and other key collectives is more than interdisciplinary—it is fundamentally *trans*disciplinary. Transdisciplinarity "does not simply mean laying two or more disciplines next to each other. Rather, it means to set about a question simultaneously taking into account visions and methods on the same topic from seemingly different perspectives" (ENSN 2008). Moreover, *transdisciplinary* also points to the significant challenges researchers may encounter that arise from profound "differences in research methods, work styles, and epistemologies" (Klein

2004, 520) when scholars from radically different academic backgrounds attempt to merge their work together.

While many of the social implications of neuroscience are being addressed by scholars in neuroethics[8] and the burgeoning field of critical neuroscience,[9] few are addressing the *scholarly* implications posed by these new transdisciplinary partnerships: the promises as well as the difficulties. Research on "the science of team science" (Stokols et al. 2008) has begun to illuminate the prospects and pitfalls for large-scale interdisciplinary research teams charged with transdisciplinary tasks.[10] One of our goals with this collection is to open similar conversations concerning research collaborations at the dawn of this most recent turn to neuroscience. In line with Daniel Stokols et al., we view transdisciplinarity as "a process in which team members representing different fields work together over extended periods to develop shared conceptual and methodologic frameworks that not only integrate but also *transcend their respective disciplinary perspectives*" (S79, emphasis added).

This volume represents a first attempt to think historically, practically, and critically about the neuroscientific turn from the perspectives of adopters—and critics—in the humanities and social sciences, but also from the perspectives of neuroscientists themselves. Through this collection, we develop a preliminary framework for defining and theorizing the neuroscientific turn by making three arguments. First, we contend that the neuroscientific turn has a longer and more dynamic history than might be evident from the hype surrounding the emergent neurodisciplines or the earlier "decade of the brain." We argue that the neuroscientific turn should be understood contextually, via precursors that date back to the nineteenth century and beyond. In part 1 of the collection, a genealogy of these antecedents challenges the novelty, and thus the very definition, of this latest "neurorevolution."

Our second—and related—argument is that neuroscience is a *translational discipline:* a set of methods and/or theories that has become transferable—sometimes problematically so—to other disciplines. We contend that the complexity of the brain and its centrality in most human endeavors ought to be reflected in equally complex, multifaceted modes of inquiry, perhaps best facilitated by transdisciplinary conversations. In part 2, practitioners from multiple fields explore the multidirectional traffic of ideas, theories, and methods between the translational discipline of neuroscience and their own fields. Each offers suggestions and directions for future research in areas of history, theology, literary studies, literacy, and neuroethics. Thus, this collection provides the opportunity for diverse fields to speak with and alongside one another; for this reason, we hope to present our vision of the ideal neuroscientific turn as a collaborative project rather than a vestigial outgrowth of the neurosciences or a return to biologism.

The specter of reductionism, essentialism, and biologism leads us to our third argument, that the challenges we face as transdisciplinary scholars are also revelatory: they illuminate theoretical assumptions and boundaries that can be just as easily reinforced as reconfigured at moments of disciplinary upheaval. Coming as we do from backgrounds in the fields of rhetoric of science and of literature and science, we actively seek to trouble disciplinary boundaries, proper objects, and the cultural and institutional hierarchies between the natural sciences, the social sciences, and the humanities. Yet, we also recognize—as do the contributors to this volume—that there are particular challenges and limitations to transdisciplinary endeavors, including the dilemmas of field-specific language (jargon) and translation; the issues of disciplinary training, gatekeeping and boundary work; and the questions of funding and access to technologies and equipment. In part 3, critical responses by neuroscientists, economists, and a literary critic explore the power, promise, and potential perils of the neuroscientific turn for disciplinarity in the twenty-first-century academy.

In this introductory essay, we provide the context for our—and the collection's—collective arguments: we provide a brief overview of neuroscience and (re)define the field as a translational discipline; we explore possible significations of the "neuro-" prefix, including its potential to black-box the complexity and historicity of neuroscience research; we contextualize the turn to neuroscience within a framework of other large- (and small-) scale academic shifts; and, finally, we provide an overview of the chapters that make up this collection.

Neuroscience as Translational Discipline: Theory and Method

Neuroscience stands alongside molecular genetics and evolutionary biology as one of the most significant interdisciplinary collaborations in the history of the life sciences. The field is mammoth and difficult to define: a conglomerate of biology, chemistry, physics, and medical sciences that produces research on the smallest glial cell to the evolution of human consciousness. Although the brain gets the most attention, it's worth remembering that neuroscience is the study of the human nervous system, which includes the spinal cord as well as elements of the peripheral nervous system: the network of tingling highways that transfers information about the external environment back to the brain.

Several key studies have explored neuroscience's various laboratory cultures as well as neuroscience's appeal as it travels in and through the public sphere (Dumit 2004; Joyce 2008; Saunders 2008). While we acknowledge the importance of following scientists, following facts, and following the media hype, our concern in this collection is with the effects of neuroscience in and on the disciplines of the academy—specifically the humanities and social sciences. In both arenas, neuroscience has significant translational appeal. In scientific contexts, the adjective *translational* means the difference between research that stays in the laboratory and research—technologies, equipment, methods—that is applicable in extralaboratory settings. In STS contexts, the noun *translation* is used for processes by which research moves outward from a specialized disciplinary field; this movement occurs when science is translated for a lay audience *and* when scholars from various disciplines attempt to speak to each other across the divides of jargon, method, and theory. We contend that neuroscience has become what we would call a *translational discipline:* a set of methods and/or theories that has become transferable to other disciplines *and* a flash point for transdisciplinary exchange.

To put it simply, neuroscience has become not only an appealing mode of investigation but also an epistemology that both transcends and reinforces thinking about the brain and mind. In the neuroscientific turn, research using tools such as fMRI reproduces assumptions about the primacy and centrality of the brain to aspects of the self and behavior; fMRI in particular purports to answer our deepest questions about personhood, knowledge production, and human behavior by concretizing each of these inquiries into technical studies of brain activity and presenting the data in images that look very much like maps to the mind. *At the same time,* the neuroscientific turn's deployment of neuroscience also "synthesizes and extends discipline-specific theories, concepts, methods, or all three to create new models and language to address a common research problem" (Stokols et al. 2008, S79). This collection represents a first attempt to encourage and capture the "new models and language" that we hope will inspire robust and multidimensional research.

According to our definition, at least three distinct groups are engaging with neuroscience as a translational discipline. First are neuroscientists who have decided to ask questions about objects and subjects that are traditionally under the purview of the social sciences and humanities. Take Robert Persinger, for example, one of the most notable figures in the field of neurotheology. By stimulating portions of his subjects' brains (with a device now known colloquially as the "god helmet"), Persinger was able to produce a feeling in his subjects "for which they employ terms that have usually been reserved for religious states and shamanic traditions" (2001, 520). Is it any wonder that Persinger, armed with evidence that human religious experience is all in our heads, feels the need to step outside of the strictures of scientific discourse to boldly attest that "the brain mechanisms and electromagnetic patterns within the brain that generate the god experience might be considered one of the most important challenges to which neuroscience must respond" (519)?

> God experiences, which are often employed as proofs of god beliefs, are likely to have been responsible for more human carnage in the history of civilization than any single pestilence. Peoples have been killed in the name of a god by others because they did not believe in the same god. If this propensity for group aggression is coupled to the same or similar processes as the god experience, then we as a species might wish to discover all of the stimuli, endogenous and exogenous, that can unleash these behaviors within a group. (514)

With world peace as a potential outcome, it's hard to find a science with more translational appeal. One of the dangers of translational research, however, is its propensity to swallow the multiplicity of perspectives that can make interdisciplinary research so effective. As Persinger notes, "From the perspective of modern neuroscience all behaviors and all experiences are created by the dynamic matrix of chemical and electromagnetic events within the human brain"—a totalizing claim that leaves little room for commentary from religious studies scholars, anthropologists, or STS scholars (2001, 515). Our hope is that the neuroscientific turn's translational appeal can also produce an interest in reflexive analysis, which may need to emerge from the second group of scholars incorporating neuroscience into their work.

This second group includes those trained in the humanities and/or social sciences that have moved out of the proverbial library and into the laboratory. Many of the adopting fields, including (but not limited to) economics, theology, and political science, have employed the actual technologies and methods of neuroscience, most notably functional magnetic resonance imaging (fMRI) and data analysis methods such as subtraction. Take, for example, neuroliterature (neurolit or neurolit criticism), which has not—until very recently, that is—been interested in scanning the brain. Instead, the field has emerged from a collaboration between

English literature, psychology, and neuroscience that relied on cognitive theory[11] as its theoretical backbone. However, the field is changing rapidly: students in the Yale-Haskins Teagle Collegium headed by Michael Holquist (a comparative literary critic) are designing fMRI experiments to ostensibly analyze brain activation during reading-activities.[12] Institutions such as the University of Pennsylvania are making such collaborations possible by offering programs like "Neuroscience Boot Camp" through which humanists and social scientists can learn more about neuroscience. According to the program's website, "The Penn Neuroscience Boot Camp is designed to give participants a basic foundation in cognitive and affective neuroscience and to equip them to be informed consumers of neuroscience research." The challenge, particularly as the neuroscientific turn progresses, will be to avoid unidirectional translations that would designate humanists and social scientists as mere "informed consumers of neuroscience research." To put it another way, our hope is that in these emerging partnerships, neuroscience need not always take the position as first author. Instead, as this collection demonstrates, humanists and social scientists are, and will continue to be, vital coproducers in the process of inquiry.[13]

This is nowhere more evident than in the third group of scholars engaged with the neuroscientific turn: humanists and social scientists who often collaborate with the neurosciences through secondary data analysis and/or by repurposing neuroscience as theory. Paul Eakin, for example, claims to have been "inspired" by the popular neuroscience texts of Antonio Damasio to approach autobiography in a "new way" (2004, 124) that incorporates insights from neuroscience concerning "neurobiological rhythms of consciousness" (130). His argument does not rely on MRI scan time, participants, or protocols; instead, he employs neuroscience as philosophy or theory to supplement and support his own arguments. Although some emergent neuroscholarship makes more effective use of neuroscientific research than others (Johnson and Littlefield 2011), there is enough evidence of the use of neuroscience as theory to warrant the naming of a general trend, one that has the potential to help (re)shape the neuroscientific turn into a collaborative conversation. The emergent neuroscholarship showcased in this collection, for example, has largely been designed and performed from translational positions outside of neuroscience's traditional disciplinary bounds.

For each group (neuroscientists who turn to philosophical questions, humanists and social scientists who turn to neuroscience as method, and humanists who turn to neuroscience as theory) translation becomes a crucial concern. Lieberman and colleagues (2003) make this argument concerning the rise of the neuroscientific in political science.

> Researchers in heretofore foreign disciplines will need to collaborate, and to do so they must learn to speak one another's specialized languages. Political scientists who wish to use neuroscience methods will have to acquire a working vocabulary of foreign concepts—including neuroanatomical terms such as prefrontal cortex and hippocampus—and will have to learn techniques such as functional magnetic resonance imaging (fMRI). The same is true for cognitive neuroscientists who wish to study political and related social-cognitive phenomena. They will have to learn about attitudes, stereotyping, political sophistication, and how to manipulate mood and motivation. It bears repeating that the benefits of these efforts are wholly practical and that each field has much to offer the other. (694)

Neuroscience may be a translational discipline, but other kinds of translations will be necessary along the way for the neuroscientific turn to surmount transdisciplinary "Babelisation" (Nicolescu 2002, 40) and become a multidirectional conversation among disciplinary equals.

The Black Box of the "Neuro"

One of the many challenges of the neuroscientific turn is definitional: because the "neuro" prefix has been so widely applied to disciplinary undertakings (neuroaesthetics, neurotheology, neuroeconomics, neuromarketing) and to other common words,[14] it has lost a specificity of definition. Is neurohistory the same application of/dialogue with neuroscience as neuropolitics? Is the "neuro" of neurotheology the "neuro" of neurology? From a structuralist perspective, neuroscience is a signifier that, in its translation(s), is often disconnected from any particular or concrete signified. Throughout the collection and in the title, we refer to recent changes as the "neuroscientific turn," rather than the "neuroturn," despite the admittedly seductive power of the latter. This is somewhat arbitrary given the ubiquity of *neuro* as prefix; however, by doing so we intend to call into question the "neuro" as a universal referent, remembering always that the "neuro" (whether it refers to the brain or to neuroscience) is a historical object, created and shaped by inquiry.

From a skeptical perspective, we could read the "neuro" as ubiquitous and diluted. As a prefix, *neuro-* vaguely references the brain, imaging technologies (CAT, PET, MRI), the medical specialty of neurology, and maybe even science fiction (e.g., the title of William Gibson's *Neuromancer* [1984]). The "neuro" signifies a hypothetical *location* (i.e., the nervous system, brain, neuron) where we should look for answers to our deepest questions about consciousness, learning, selfhood, and so forth. It does not acknowledge that this location may be a chimera—one that does not necessarily correspond to (states of) mind; nor does it account for the degree and variety of translations and representations that construct the brain as an accessible locus of information/data (Wilson 1998; Beaulieu 2001, 2004; Joyce 2008).

This is not to suggest that applications of the "neuro" are uniformly uncritical. Many, including those found in science fiction and in several of the disciplinary applications, serve to complicate our understanding of the "neuro" as either a location of or source for transparent information. Some disciplinary adopters, including sociology and anthropology, have expressed skepticism about taking on the methodologies of the neurosciences. They view neuroscience as a collaborator and equate their disciplinary combination to a "marriage" (Northoff 2004, 748) in which both qualitative and quantitative methods can work in tandem. Leaving aside the usefulness of the marriage metaphor (after all, the history of the institution can hardly be said to be one of collaborative equals), Northoff believes that "the encounter with neuroscience will allow anthropology to map its own boundaries and thus, where the two disciplines intersect" (2004, 750). This gatekeeping language implies a critical-practical perspective on the "neuro" as useful but not necessarily engulfing.

On the second count, and in a more theoretical vein, we could also read the "neuro" as a kind of Latourian black box that stands in for "neuroscience"; in this case, the *neuro-* prefix becomes a disciplinary perspective/technology whose origins, inner workings, and possible challenges have been replaced by a facade of the always-already. While neuroscience remains a dynamic enterprise, we often fail to recognize its genealogy, acknowledge its fallibility, or probe the limits of its applications. The best we can do is reopen the box or study the controversies central to its construction (Latour 1987). Work by Lisa Cartwright (1992), Anne Beaulieu (2004), Andrew Shail and Laura Salisbury (2010), and Fernando Vidal (2009) has taught us that if we reopen the black box of neuroscience at particular instances, we may see that the collaborative science has a longer, more punctuated history. Our contributors explore three historical angles in part 1 of the collection; here we want to examine—and open—the black box of neuroscience in several instances.

Any discussion of the recent proliferation of the "neuro" must include a comment on the expansion of interest in neuroscience during the 1990s, famously declared by U.S. president George H. W. Bush to be the "decade of the brain" (DoB). The purpose of Bush's proclamation was to encourage public support for neuroscience research and to stimulate both academic and industry interest in the brain. Indeed, each year during the 1990s, the Society for Neuroscience's numbers swelled by about a thousand (Jones and Mendell 1999, 739); and according to Antonio Damasio, during the 1990s, "more may have been learned about the brain and mind . . . than during the entire previous history of psychology and neuroscience" (1999, 112). The DoB was not simply about enrolling academic actors in the study of the human brain; it also aimed to stimulate political support for neuroscience, resulting in a rather strange Latourian "alliance" of politicians, NIH administrators, scientists, and disease advocacy groups all clamoring for more neuroscience research (Jones and Mendell 1999). Although the DoB was not the *origin* of the recent neuroscientific turn, the DoB undoubtedly served to organize and intensify discourses of the brain and mind that, in turn, have helped to legitimate the turn to neuroscience in fields from outside the life sciences.[15]

Yet for all its influence in recent years, both on the academy and within the public imagination, the DoB belongs in a longer genealogy that better historically situates the neurosciences, one that begins well before 1990. One of the challenges is that the neurosciences and those working within the neuroscientific turn alike often eschew history in order to appear new and relevant. In his work on "brainhood," Fernando Vidal argues that discourses of newness and technophilia are "typical of the ahistorical triumphalism characteristic of the neuro field" (2009, 10); he maintains that "neurocultural discourses, and neuroethics with them, mask the continuity that exists, since the early nineteenth century, in the main assumptions, in the "big" questions being asked (about the nature of consciousness or the mind-brain relation), and in the answers to them as well (e.g., mind as reducible to brain or mind as an emergent property)" (2009, 10). Here, the "neuro" of neuroethics functions to make long-standing queries about ethics appear novel and in need of fresh intervention (a critique that Rosenquist and Rothschild also raise about neuroeconomics in this volume).

To clarify Vidal's—and our—point, we should note that Adeline Roskies has argued that the burgeoning field of neuroethics comprises two different branches (2002): the "ethics of neuroscience," which applies traditional bioethical concerns about experiment design and the use of human subjects to the new neurotechnology; and the "neuroscience of ethics," which applies new findings in neuroscience to our thinking about human moral agency. The latter fits more squarely within the rubric of the neuroscientific turn; the former's classification proves to be a bit more complicated. On one hand, the ethics of neuroscience has played a crucial role in our understanding of the pitfalls of neuroscience; on the other hand, the ethics of neuroscience has helped create the very objects it promises to manage.

Instead of vilifying the ethics of neuroscience, we would argue that it is crucial to note that both branches of neuroethics are partners and participants in the larger neuroscientific turn—a shift that has (re)inforced the ahistoricism of neuroscience. As with Ulrich Beck's "risk society" (1992) in which science creates the risks it manages, we could argue that the neuroscientific turn—and not simply the neuroscience of ethics—supports "those who benefit from the assumption that we are cerebral subjects, and claim that the assumption rests on neuroscientific discoveries" (Vidal 2009, 8). By creating a field, proper objects, and a manifest purpose for themselves, scholars engaged in the neuroscientific turn are no different from any other academic group. We are, in short, all guilty of a certain kind of risk-management construction that must be historicized.

The new field of critical neuroscience, developed by Suparna Choudhury, Saskia Nagel, and Jan Slaby, is currently engaged in paying these same questions forward by inquiring into the purview, impact, and future of the neurosciences. Scholars working within this rubric seek to intervene in the neurosciences and their popular representations by providing critical interpretations: "to develop a more thorough self-understanding and awareness of the social implications of research and its uses, among neuroscientific practitioners, which can then be fed back into the practice of neuroscience" (2009, 62). And their manifesto is explicit about exploring "the historical contextualization of the development of the neurosciences" alongside

> the study of socio-economic drivers of research programmes; ethnographic analysis of laboratory practices, conceptual and technical scrutiny of methodologies, including the paradigms, data analysis and interpretive frameworks; media representations and their interaction with processes of research; engagement with the mechanisms of agenda setting in the neurosciences; and, most importantly, the implementation of alternative approaches, methods, designs and interpretations in neuroscientific research. (2009, 62)

Critical neuroscience promises to produce important interventions both in theorizations of popular representations and in social policy; however, the purview of their analyses does not turn a mirror back onto the academy and its adoption of the "neuro" as prefix, which is the particular project of this collection.

One final—but by no means the only other—way to open the black box of neuroscience and its translational prefix, "neuro," is to broaden our scope and definition of what counts as precursory. Take, for example, William James's concern with what he calls "medical materialism" at the turn of the twentieth century. In lecture one of *Varieties of Religious Experience* (1902), James discusses "Religion and Neurology."[16] While there are some key distinctions between neurology and neuroscience[17] in contemporary scientific discourse, these distinctions did not exist at the turn of the twentieth century. The differences that mattered between philosophy, psychology, physiology, and neurology—the four fields concerned with questions of the mind/brain at the turn of the twentieth century—were often resolved on the side of the natural sciences (physiology and neurology). James, a proponent of early psychophysiological experimentation, theorized an earlier, neurological turn, calling it "medical materialism." In this particular lecture, he addresses those who have attempted to explain religious experiences (e.g., trance) exclusively via the body—and, for our purposes, occasionally the brain.

> Medical materialism seems indeed a good appellation for the too simple-minded system of thought which we are considering. Medical materialism finishes up Saint Paul by calling his vision on the road to Damascus a discharging lesion of the occipital cortex, he being an epileptic. It snuffs out Saint Teresa as an hysteric, Saint Francis of Assisi as an hereditary degenerate. . . . And medical materialism then thinks that the spiritual authority of all such personages is successfully undermined. (1936, 14)

If we were to compare James's list to the work of neurotheologists, such as Persinger, we would find his criticism remarkably contemporary. However, unlike some of the neurotheological work of late, James did not wish to explain away religious experience via neurology. Instead, he espoused a pragmatic approach in which religious experience could be explained in tandem and comparison with other emotional states. "Religious happiness is happiness," maintains James; "religious trance is trance."

> And the moment we renounce the absurd notion that a thing is exploded away as soon as it is classed with others, or its origin is shown; the moment we agree to stand by experimental results and inner quality, in judging of values, who does not see that we are likely to ascertain

the distinctive significance of religious melancholy and happiness, or of religious trances, far better by comparing them as conscientiously as we can with other varieties of melancholy, happiness, and trance, than by refusing to consider their place in any more general series, and treating them as if they were outside of nature's order altogether? (1936, 25)

In his entreaty to "renounce the absurd notion that a thing is exploded away as soon as it is classed with others," James maintains the power of experientialism, while also increasing the authority and influence of medical materialism; in so doing, he redefines the relationship between the two. No longer antagonistic by nature, scientific explanations of religious phenomena are redefined as classifications.

We have taken some pains with James's "Religion and Neurology" because it widens the genealogy of the neuroscientific turn. James's critique of medical materialism, particularly given its links to neurology, represents an earlier move to classify religious experiences as brain events.[18] While Saint Paul's reclassified visions were not under the auspices of neurotheology, they could very well be revisited under that rubric today. Indeed, one of the essays in this volume (Hendrix and May) represents neurotheological research in this vein. Instead of thinking of the neuroscientific turn as something new and revolutionary, an opening of the "neuro"/neuroscience black box reframes the turn as historically dynamic and illustrates how particular aspects (such as the propensity to accept the authority of biological explanations and reclassifications) develop and change over time.

Turn, Turn, Turn

Given the number of "turns" in humanities scholarship, and a similar—but less frequent—set of turns in the social sciences, a neuroscientific turn may seem like only the latest interdisciplinary fad.[19] And in at least one sense it is: as with previous turns, it may be that some scholars from the humanities and social sciences are adapting to an increasingly difficult financial climate by adopting other disciplinary theories and methods as a means to survive and remain relevant in an academic world that privileges scientific research, and particularly scientific research with translational benefits.[20] In this adaptative sense, the neuroscientific turn may represent the outcome of a struggle to ensure one's reproductive fitness in a hostile environment, so to speak. However, there are several additional and more positive ways to interpret the turn.

William Connolly has commented that the turn to scientific methods in the humanities and social sciences may offer a corrective to an anti-science reductionism of cultural theorists: "In their laudable attempt to ward off one type of reductionism too many cultural theorists fall into another: they lapse into a reductionism that ignores how biology is mixed into thinking and culture and how other aspects of nature are folded into both" (2002, 2). From this perspective, the neuroscientific turn aligns itself with the bodily and material(ist) turns that created the opportunity for revisiting the essentialist/constructivist debates of the previous decades, and opened up productive areas of research in feminist theory and disability studies, among others.

In its multidirectionality, *The Neuroscientific Turn* is both a return to and a complication of the "two cultures" debate that has animated disciplinary relations between the humanities and sciences for over half a century. While challenging the hierarchies that prioritize science, this collection also creates a forum for scientists, humanists, and social scientists (the "third culture") to think, theorize, question, and reframe the neuroscientific turn *in vivo*, defining it at the outset as a dynamic, dialogic enterprise. We suggest that the neuroscientific turn can be (re)imagined as multidirectional, given that the humanities and social sciences are not merely adopting or responding to the methods and theories of the neurosciences, but are—at least in some cases—engaging in bi-directional conversation. Consider these remarks from a recent essay concerning neuroeconomics.

> Although we focused solely on applications of neuroscience to economics, intellectual trade could also flow in the opposite direction. Neuroscience is shot through with familiar economic language—delegation, division of labor, constraint, coordination, executive function—but these concepts are not formalized in neuroscience as they are in economics. There is no overall theory of how the brain allocates resources that are essentially fixed (e.g., blood flow and attention). An "economic" model of the brain could help here. Simple economic concepts, like mechanisms for rationing under scarcity, and general versus partial equilibrium responses to shocks, could help neuroscientists understand how the entire brain interacts. (Camerer, Loewenstein, and Prelec 2005, 56)

Here, a multidirectional perspective illuminates not only a history of unrecognized discursive interactions between neuroscience and economics but also the conviction that the former could benefit from the latter's expertise.

Metaphorically, the potential dynamism of the neuroscientific turn could be theorized using one of the most striking findings that emerged during the decade of the brain: neuroplasticity[21] (also called "cortical plasticity," and "cortical remapping"). For years it was a truism that the brain was a static entity whose structure was fixed after childhood (a belief that made its way into the popular imagination as the conviction that teenage brain cells could be permanently "fried" by drug use). In recent years, however, neuroscientists have argued that thinking may physically change the brain's structure, providing physical evidence of Freud's law of association by simultaneity and giving rise to the neuroplasticity mantra that "neurons that fire together wire together." While we do not champion or challenge neuroplasticity in this volume, we do find it a useful metaphor. Indeed, the principle of neuroplasticity has potentially significant clinical applications (in treating patients with brain trauma, stroke, or obsessive-compulsive disorder), but it also gives rise to a new way of thinking about two old ideas: the interaction of biology and culture, and the division of mind from brain.[22]

The idea that culture can shape biology has been made powerfully by Nicholas Carr's trade book *The Shallows* (2010), which draws on neuroplasticity research to argue that the Internet is literally changing the brains of its users. William Connolly has argued that findings like those referenced by Carr necessitate a new politics that acknowledges the complex means "through which cultural life mixes into the composition of body/brain processes. And vice versa. The new neuroscience, while needing augmentation from cultural theory, encourages students of culture to attend to the layered character of thinking; it also alerts us to the critical significance of technique in thinking, ethics, and politics" (2002, xiii). In essence, Connolly argues that what is needed is a hybrid form of scholarship that brings neuroscience and cultural theory together as partners in a joint effort. Conceived as a partnership between the arts, sciences, and cultural criticism, a field like neuroaesthetics, for example, not only needs to examine the structures in the brain that allow us to perceive the beautiful but also might ask how art might alter the human brain; how literary movements have literally changed the way that we think; how art, politics, and the body intermingle in productive and unexpected ways. Semir Zeki, one of the pioneers of neuroaesthetics, has argued that such a field is needed because art is the product of the human brain (1998, 78); neuroplasticity offers the tantalizing possibility that the brain is also the product of the art it consumes. This kind of thinking will help produce a neuroscientific turn that is not merely reactive but productive.

In this vein, we need only think of the strides made by earlier turns, such as the linguistic turn (Rorty 1967), which introduced the idea that language structures thought and disciplinarity. We contend that the neuroscientific turn has required and continues to require the thoughtful invention of a multidisciplinary, interpretive language. As has been the case with other imaging technologies (from cinema to the X-ray to Positron Emissions Tomography—PET), those technologies employed by the neurosciences and used in the neuroscientific turn (MRI, fMRI, EEG) are products and producers of new ways of seeing. These new ways of seeing have been historically cataloged by numerous scholars;[23] theorizing the neuroscientific turn along similar lines augments our knowledge of the ways that technologies of visualization require translation. Translation involves first and foremost a certain kind of tolerance and an appreciation for the rigorous training endured by academics *of all disciplines.* Humanists and social scientists need to better familiarize themselves with the jargon and the scientific methods of imaging; scientists need to better value the descriptive—and theoretically challenging—work typically associated with the nonscience disciplines (cf. Lieberman et al. 2003). Translation also involves the recognition that imaging technologies can only show us so much. As most imaging technicians explain, fMRI and other scanning technologies produce data that is then passed through several interpretive rubrics, a point explored in detail in this collection by Susan Fitzpatrick. As a result of some basic communication improvement, the neuroscientific turn can produce a new vocabulary that is less vague than the "neuro" that, so far, dominates discussion from all sides.

Transdisciplinary Reading: A Primer

In the spirit of good communication, we would offer a brief suggestion about how to read this book. While it may be tempting to flip immediately to the essays from authors from your home discipline, we ask that the reader approach the essays with the same transdisciplinary spirit that inspired this book. Although we have asked each author to translate their work for a generalist audience, we ask that the reader be mindful that the authors in our collection come from a variety of fields, each with its own evidentiary norms, theoretical assumptions, and critical vocabulary. We strongly believe that there is value in trying on another's theoretical perspective, even if it's not always the perfect fit.

Likewise, we have encouraged contributors to engage in active dialogue with one another via a Wiki site that allowed authors to read and reference each other's essays. In several locations throughout this collection, including a response essay (by Anne Beaulieu) and an afterword (by

Joseph Dumit), contributors address and engage with the challenges and controversies that arise when one reads transdisciplinarily. At the cusp of this neuroscientific turn, we believe this mixture of critical questions and intellectual generosity is and will remain the best recipe for any transdisciplinary endeavor.

Arguments and Contributors

Part 1: The Neuroscientific Turn in Context

We contend that the neuroscientific turn is historical; the recent adoption of neuroscience by multiple disciplines is not a novel phenomenon but one with a distinct history. Our genealogical reading, which began in this introduction, is taken up by three of our contributors so as to situate the neuroscientific turn as one of many moments when the brain sciences precipitated disciplinary strife, consolidation, and revaluation. Instead of thinking of the neuroscientific turn as something novel and out of the ordinary, we propose looking backward (and sideways) at similar moments; in this respect, the neuroscientific turn will not seem quite so "revolutionary." While some may ask "why is the neuroscientific turn happening in the academy *now*," we would ask "why is the neuroscientific turn happening in the academy *again*." In thinking historically about the neuroscientific turn, we glean instructive lessons about collaborative, multidisciplinary thinking. In the first part of the collection, then, contributors examine the elements, trends, and contexts for the most recent of several neuroscientific turns.

Essays by Justine S. Murison, Jameson Kismet Bell, and Kélina Gotman each look to the past as a means to understand the present moment not as a revolution but as a reiteration. Murison argues that the examination of an earlier turn, the neuroscientific turn of the 1840s, illustrates how the science of neurology and various popular literatures worked in (uncoordinated) conjunction to alter the language of selfhood. Kismet Bell's essay explores a set of events in sixteenth-century Strassburg in order to recontextualize the brain, which was stripped of its connections to theological, scholastic, literary, and mystical signs. His argument reminds us of the ways that disciplinary language has created a particular—and largely fictional—brain. Gotman's essay explores how neuroscience animated the development of post-poststructuralist philosophy. In particular, she illustrates the imbrications of modernist and nineteenth-century notions about the singular and the plural that have been captured by the neurosciences and what she calls the "neural metaphor."

Part 2: The Neuroscientific Turn in Practice

Ours is not a collection of polemical critiques or a forum for gatekeeping and boundary work. Instead, and from the outset, we intended to collect various perspectives on the neuroscientific turn from practitioners in the social sciences, humanities, and neurosciences and place these alongside contextual, historical, theoretical, and methodological essays so as to broaden the scope of the neuroscientific turn as it has emerged thus far. In part 2, "The Neuroscientific Turn in Practice," essays by Sarah Birge, Scott E. Hendrix and Christopher J. May, Gwen Gorzelsky, and Eric Racine and Emma Zimmerman explain how the neuroscientific turn has the potential to create new knowledge and reveal how disciplinary objects are formed. Birge's essay explores the multidirectional traffic between the humanities and neuroscience by analyzing neuroscience's engagement with "neurofiction" by Richard Powers, Mark Haddon, and Jonathan Lethem, among others. In one of the more forward-looking collaborations included in the collection, Hendrix and May bring together expertise in history and neuroscience to create a neurohistorical account of two mystics, Bernard of Clairvaux (twelfth century) and Teresa of Ávila (sixteenth century). In her essay on intersections between neuroscience and literacy, Gorzelsky argues that a survey of literate practices concerning meditation can reveal new areas for neuroscientific research while simultaneously alerting researchers to areas in which neuroscience can be supplemented by ethnographic analyses. Finally, Eric Racine and Emma Zimmerman explore the challenges of bringing together ethics and neuroscience. Using an approach informed by John Dewey and other pragmatists, Racine and Zimmerman chart the dangerous waters of neuroessentialism, neural determinism, and neurorealism. As an illustration of the dialogues we believe this collection can and will create, we include a response essay by science and technology scholar Anne Beaulieu. Beaulieu's work on neuroscience and imaging has been instrumental to the theorization of functional magnetic resonance imaging (fMRI), among other technologies. Here, Beaulieu focuses on some critical questions for the transdisciplinary work performed in the name of the neuroscientific turn.

Part 3: Critical Responses to the Neuroscientific Turn

After considering historical contexts and transdisciplinary collaborations, the collection focuses, finally, on several challenges for the neuroscientific turn. In the final part of the collection, essays by James Niels Rosenquist and Casey Rothschild, Susan M. Fitzpatrick, Peter J. Whitehouse, and Bruce Michelson offer critiques that champion transdisciplinary possibilities as much as they challenge the uncritical adoption of the neurosciences by other fields. Rosenquist and Rothschild explore neuroeconomics as a descriptive and prescriptive endeavor. In the case of the former, neuroscience may offer particular insights into behavior; in the case of the latter, the authors are more skeptical given concerns about neuroessentialism. Along with Christopher J. May, Susan Fitzpatrick and Peter Whitehouse serve as our resident neuroscientists. In his partially autobiographical essay, Whitehouse argues for a tempering of neuroethics. Drawing from examples from his long career as a clinical and academic neuroscientist, Whitehouse grounds his critiques in his work with Alzheimer's patients. Like Whitehouse, Fitzpatrick expresses skepticism about the neuroscientific turn by providing a primer in neuroscientific imaging. Her essay explores the desirability and dangers of scanning via what she has termed the problem of the "cognitive paparazzi." If the preceding essays are skeptical, Michelson's is cautiously optimistic. His essay, an in-depth explanation of interdisciplinary pedagogical practice concerning a course on the mind sciences offered by an English department, provides a central thread that blends contextual material, ideas for practice, and a critical response to the neuroscientific turn. To close the volume, Joseph Dumit makes a strong case for the importance of careful, critical engagements with neuroscience literature. There are significant risks that accompany uncritical consumption of scientific research, Dumit argues, and they are much more than just an issue of scholarly import. Claims linking brain activity and social activity in particular are finding currency in courtrooms and in policy discussions, deliberative sites where—as history teaches us again and again—the uncritical adoption of scientific claims about human behavior can have disastrous consequences. Taken together, these final essays join the collection's larger conversation in complicating this neuroscientific turn by thinking broadly and in terms of contexts, practice, and critique.

NOTES

1. The application of and reliance on neuroscience within medicine continues to grow. As Kelly Joyce notes in *Magnetic Appeal* (2008), actual figures are difficult—if not impossible—to obtain (110–13). What we do know, thanks to several sources such as the Information Means Value and the Centers for Medicaid and Medicare Services, is that the number of MRIs is increasing, extending a trend that began in the 1990s (Joyce 2008, 110–13).

2. Because the neuroscientific turn is a relatively recent phenomenon, emer-

gent neurodisciplines remain, thus far, either connected to their parent disciplines or clustered as neuroscientific subfields. This is to say that few, if any, emergent neurodisciplines have established their own journals or academic societies. Neuroeconomics is a notable exception to this trend, having established the Society for Neuroeconomics in 2010. The remaining neurodisciplines have been addressed in mainstream journals; take, for example, Jordynn Jack and Laurence Appelbaum's special issue on neurorhetoric that was published in *Rhetoric Society Quarterly* (2011) or Jeff Pruchnic's use of the term *neurorhetoric* in *Configurations* (2009). As early as 1997, Warren TenHouten published an essay on "Neurosociology" in the *Journal of Social and Evolutionary Systems*. In addition to these types of publications, scholars working in many of the emergent neurodisciplines have published academic and trade books about their subjects; these are too numerous to list here.

3. While there is a certain critical mass behind the neuroscientific turn in the academy, it is too early to make claims about which disciplines have more readily taken to neuroscience. In part, this is because disciplines don't make decisions, they shift and change with the collective intelligence of a swarm: they reflect tendencies and patterns, and it is often difficult to attribute or even speculate on causality. We would note that in fields that must regularly adjudicate about the latest scientific evidence, as in law (or neurolaw), the neurosciences are being introduced to courtrooms and integrated in legal precedents, prompting many conversations about admissibility, conversations reminiscent of *Frye v. U.S.* (1923) and *Daubert v. Merrell Dow* (1993).

4. See also V. S. Ramachandran's 5th Reich Lecture, BBC Radio 2003, "Neuroscience—The New Philosophy."

5. The phrase *neuroturn* or *neuroscientific turn* has slowly proliferated in scholarship from multiple disciplines, as we will detail here. However, these references have not yet cohered into the kind of focused inquiry we provide in this collection. First, and perhaps most important, Andreas Roepsdorff (Aarhus University) used the term *neuroturn* in his January 2007 lecture, "The Neuroturn: Challenging Anthropology or Anthropological Challenge?" We found Roepsdorff's lecture after conceiving of this volume; however, and given its affinity with our own use, we want to credit Roepsdorff with recognition of the turn as early as 2007. Catherine M. Lord refers to the "'neuroscientific' turn" once in her 2009 paper "Angels with Nanotech Wings: Magic, Medicine and Technology in Aronofsky's *The Fountain*, Gibson's *The Neuromancer* and Slonczewski's *Brain Plague*." In 2005, Prisca Augustyn used the phrase *cognitive/neuroscientific turn* in her paper "Art—Depression—Fiction: A Variation on René Thom's Three Important Kinds of Human Activity." Finally, the phrase *neuroscientific turn* has been used in reference to the Decade of the Brain (Wilson 2004).

6. Take, for example, the Center for Integrative Neuroscience and MIND Lab run by Andreas Roepsdorff in Aarhus, Denmark, and the Yale-Haskins Teagle Collegium hosted by Yale. See also the work of centers and/or interdisciplinary programs such as UC Irvine's Interdepartmental Neuroscience program, http://www.inp.uci.edu/; the new Neuropolitics Center at Emory University (est. 2008); and the Neuroscience Institute at Princeton. Finally, consider Nikolas Rose's group the "Brain, Self, Society Project," the Law and Neuroscience Project, the University of Pennsylvania's Neuroscience "Boot Camps," and the 2011 summer seminar at

Stanford's Center for Advanced Study in the Behavioral Sciences devoted to neuroscience and the humanities.

7. The editors have both participated in and won funding from ENSN Neuroschools. For an overview of the ENSN Neuroschools, see Frazzetto (2011).

8. For example, Jonathan Marks (2001) has suggested that the translation of neuroscience research might be best be approached with "*neuroskepticism*—that is, a perspective informed by science studies scholarship that views with some healthy skepticism claims about the practical implications and real-world applications of recent developments in neuroscience" (4). Marks discusses the recent application of neuroscience in national security policy as a translation that demands "some difficult questions that explore the kinds of neuroscience research that are being funded and address the broader context in which that research takes place" (11).

9. The project of "critical neuroscience" is spearheaded by Suparna Choudhury, Saskia Nagel, and Jan Slaby; we will discuss their work later in the introduction. As this book was going to press, we had the opportunity to meet with these and other Critical Neuroscience Scholars at the "Neuro-Reality Check" workshop sponsored by the Max Planck Institute for the History of Science in Berlin (Dec. 2011).

10. Research on the "science of team science" and transdisciplinary initiatives (which have been funded by the NIH and other large-scale funding agencies) can be found in the *American Journal of Preventative Medicine's* Supplement 1 (2008). This issue richly contextualizes much of the relevant literature, current theoretical trends, and prospective training pedagogies for transdisciplinary research in the sciences.

11. Literary criticism turned to cognitive science and theory during the 1990s, and the two have shared an extended relationship that now partly informs neuroliterary criticism. For early work, see Alan Richardson's "Cognitive Science and the Future of Literary Studies" (1999). Cognitive theory programs are still in existence; however, neuroscience has taken much of the limelight.

12. More information on the Teagle Collegium is available via their website, http://www.haskins.yale.edu/Teagle.html. A sample media report about the Collegium and neuro lit-crit in general is available via *The Guardian*, April 11, 2010: Paul Harris and Alison Flood, "Literary Critics Scan the Brain to Find Out Why We Love to Read," http://www.guardian.co.uk/science/2010/apr/11/brain-scans-probe-books-imagination (August 11, 2010), and the *New York Times*, http://www.nytimes.com/2010/04/01/books/01lit.html.

13. The Center for Advanced Study in the Behavioral Sciences at Stanford University hosted a summer seminar on cognitive science and neuroscience in the summer of 2011 geared primarily toward humanists. One promising development: organizers Stephen Kosslyn, Anne Harrington, and John Onians plan to include not only discussions about how neuroscience might be put to use in a humanities context but also conversations about the difficulties and the limits of these collaborations.

14. A short list of terms would include *neurorevolution, neurosociety, neurodiversity, neurorealism, neuroplasticity, neuroessentialism, neurofitness, neuroculture,* and *neurotypical* (used to describe autism/Asperger's syndrome but also employed by Margaret Atwood in *Oryx and Crake* [2004, 194, 203]). Within and outside of the

academy, the neuroscientific turn has also created novums that are also of use and interest: brainhood, mind hacking, and brain customization, to name but a few.

15. The success of the decade of the brain has led to informal moves to extend the decade into the "century" of the brain (Vidal 2009, 7) as well as a recent call in *Science* for a "decade of the mind" (DoM) initiative, which proposes to build on the decade of the brain and expand its reach (Albus et al. 2007).

16. Thanks to Robert Glenn Howard for pointing us to James's discussion of medical materialism.

17. According to the *Oxford English Dictionary, neuroscience* was coined in the 1960/70s.

18. A similar response can be found in William Connolly's "Experience and Experiment" when he advocates for a "new dialogue between advocates of a science of society and those of cultural interpretation. The most promising route, in my judgment, is to forge links between neuroscience—the observational and experimental study of body-brain processes—and phenomenology, understood as the explication of implicit structures of experience that infuse perception, desire, and culture" (2006, 67).

19. Several humanistic turns include the computational turn (digital humanities); bodily turn (body studies), which is sometimes associated with the material(ist) turn; postmodern turn; pictorial turn (associated with W. J. T. Mitchell [1992], new collection out from Neal Curtis [2010]); affective and/or emotional turn; cultural turn (cultural studies); cognitive turn; social turn; and linguistic turn. In the social sciences, there has been a narrative turn, a cultural turn, a differential theoretical turn, a digital turn (all of these from sociology), and a relational turn (economics).

20. The neuroscientific turn is only one example: literature, for instance, has seen a rise in "Darwinian" literary criticism, and there is a movement for "biohistory"—the study of the impact of evolution on the course of human events.

21. The term *plasticity* is often attributed to William James in the chapter on habit in his *Principles of Psychology, vol. 1.* "Organic matter, especially nervous tissue," James surmised, "seems endowed with a very extraordinary degree of plasticity of this sort; so that we may without hesitation lay down as our first proposition the following, that the phenomena of habit in living beings are due to the plasticity of the organic materials of which their bodies are composed" (James 2007, 105).

22. Neuroplasticity has been a big hit in the popular neuroscience market (Schwartz and Beyette 1997; Schwartz and Begley 2003; Doidge 2007) and also has provided neurophysiological support for theories of cognitive behavioral therapy.

23. For a small representative sample, one might look to scholars such as Otniel Dror (1999, 2001a, 2001b), Lisa Cartwright (1992, 1995), Joseph Dumit (2004), Anne Beaulieu (2001, 2002, 2003, 2004), and Jonathan Crary (1992, 2001).

WORKS CITED

Abi-Rached, Joelle M. 2008. "The Implications of the New Brain Sciences. The 'Decade of the Brain' Is Over but Its Effects Are Now Becoming Visible as

Neuropolitics and Neuroethics, and in the Emergence of Neuroeconomies." *EMBO Reports* 9, no. 12: 1158–62. doi:10.1038/embor.2008.211.

Albus, J. S., G. A. Bekey, J. H. Holland, N. G. Kanwisher, J. L. Krichmar, M. Mishkin, D. S. Modha, M. E. Raichle, G. M. Shepherd, and G. Tononi. 2007. "A Proposal for a Decade of the Mind Initiative." *Science* 317, no. 5843 (9): 1321b–21b. doi:10.1126/science.317.5843.1321b.

Atwood, Margaret. 2004. *Oryx and Crake.* New York: Anchor.

Augustyn, Priscia. 2005. "Art—Depression—Fiction: A Variation on René Thom's Three Important Kinds of Human Activity." *Semiotica* 1, no. 4: 197–209.

Beaulieu, Anne. 2001. "Voxels in the Brain: Neuroscience, Informatics, and Changing Notions of Objectivity." *Social Studies of Science* 31, no. 5: 635–80.

Beaulieu, Anne. 2002. "Images Are Not the (Only) Truth: Brain Mapping, Visual Knowledge and Iconoclasm." *Science, Technology & Human Values* 27, no. 1: 53–86.

Beaulieu, Anne. 2003. "Brains, Maps and the New Territory of Psychology." *Theory and Psychology* 13, no. 4: 561–68.

Beaulieu, Anne. 2004. "From Brainbank to Database: The Informational Turn in the Study of the Brain." *Studies in the History and Philosophy of Science* 35, no. 2: 367–90.

Carey, Benedict. 2009. "Brain Researchers Open Door to Editing Memory." *New York Times* April 5. http://www.nytimes.com/2009/04/06/health/research/06brain.html.

Carr, Nicholas. 2010. *The Shallows: What the Internet Is Doing to Our Brains.* New York: W. W. Norton.

Cartwright, Lisa. 1992. "'Experiments of Destruction': Cinematic Inscriptions of Physiology." *Representations* 40:129–54.

Cartwright, Lisa. 1995. *Screening the Body: Tracing Medicine's Visual Culture.* Minneapolis: University of Minnesota Press.

Choudhury, Suparna, Saskia Kathi Nagel, and Jan Slaby. 2009. "Critical Neuroscience: Linking Neuroscience and Society Through Critical Practice." *BioSocieties* 4: 61–77. doi:10.1017/S1745855209006437.

Connolly, William E. 2002. *Neuropolitics: Thinking, Culture, Speed.* Minneapolis: University of Minnesota Press.

Connolly, William E. 2006. "Experience and Experiment." *Daedalus* 135, no. 3: 67–75. doi:10.1162/daed.2006.135.3.6.

Crary, Jonathan. 1992. *Techniques of the Observer: On Vision and Modernity in the 19th Century.* Cambridge: MIT Press.

Crary, Jonathan. 2001. *Suspensions of Perception: Attention, Spectacle, and Modern Culture.* Cambridge: MIT Press.

Curtis, Neal, ed. 2010. *The Pictorial Turn.* New York and London: Routledge.

Damasio, Antonio. 1999. "How the Brain Creates the Mind." *Scientific American Special Edition, "The Hidden Mind"* (December): 112–17.

Doidge, Norman. 2007. *The Brain That Changes Itself: Stories of Personal Triumph from the Frontiers of Brain Science.* New York: Penguin (Non-Classics).

Dror, Otniel. 1999. "The Scientific Image of Emotion: Experience and Technologies of Inscription." *Configurations* 7:355–401.

Dror, Otniel. 2001a. "Counting the Affects: Discoursing in Numbers." *Social Research* 68:357–78.

Dror, Otniel. 2001b. "Techniques of the Brain and the Paradox of Emotions, 1880–1930." *Science in Context* 14:643–60.

Dumit, Joseph. 2004. *Picturing Personhood: Brain Scans and Biomedical Identity*. Princeton: Princeton University Press.

Eakin, John Paul. 2004. "What Are We Reading When We Read Autobiography?" *Narrative* 12, no. 2: 121–32.

ENSN. 2008. "Neuroschools." Access date November 11, 2011. http://www.neuro societies.eu/Neuroschools/neuroschools_main.htm.

Frank, Lone. 2009. *Mindfield: How Brain Science Is Changing Our World*. Oxford: Oneworld Publications.

Frazzetto, Giovanni. 2011. "Teaching How to Bridge Neuroscience, Society, and Culture." *PLoS Biology* 9, no. 10. doi:10.1371/journal.pbio.1001178.

Harrell, Eben. 2010. "Fighting Crime by Reading Minds." *Time*. August 7. Available at *http://www.time.com/time/health/article/0,8599,2009131,00.html*.

James, William. 1936 [1902]. *The Varieties of Religious Experience: A Study in Human Nature*. New York: Random House.

James, William. 2007. *The Principles of Psychology. Vol. 1*. 1st ed. New York: Cosimo Classics.

Johnson, Jenell M., and Melissa M. Littlefield. 2011. "Lost and Found in Translation: Popular Neuroscience and the Emergent Neurodisciplines." *Sociological Reflections on the Neurosciences (Advances in Medical Sociology* vol. 13), ed. Martyn Pickersgill and Ira van Keulen, 279–98. London: Emerald Insight.

Jones, Edward G., and Lorne M. Mendell. 1999. "Assessing the Decade of the Brain." Science 284, no. 5415: 739. doi: 10.1126/science.284.5415.739.

Joyce, Kelly A. 2008. *Magnetic Appeal: MRI and the Myth of Transparency*. Ithaca: Cornell University Press.

Klein, Julie Thompson. 2004. "Prospects for Transdisciplinarity." *Futures* 36, no. 4: 515–26. doi:10.1016/j.futures.2003.10.007.

Latour, Bruno. 1987. *Science in Action: How to Follow Scientists and Engineers through Society*. Cambridge: Harvard University Press.

Lieberman, M. D., D. Shreiber, and O. Ochsner. 2003. "Is Political Cognition Like Riding a Bike? How Cognitive Neuroscience Can Inform Research on Political Thinking." *Political Psychology* 24, no. 4: 681–704.

Lord, Catherine. 2009. "Angels with Nanotech Wings: Magic, Medicine and Technology in Aronofsky's *The Fountain*, Gibson's *The Neuromancer* and Slonczewski's *Brain Plague*." *Nebula* 6, no. 4: 162–74.

Lynch, Zack. 2009. *Neurorevolution: How Brain Science Is Changing Our World*. New York: St. Martin's Press.

Marks, Jonathan. 2010. "A Neuroskeptic's Guide to Neuroethics and National Security." *AJOB Neuroscience* 1, no. 2: 4–12.

Mitchell, W. J. T. 1992. "The Pictorial Turn," *Artforum*, March 30, no. 7: 89–94.

Nicolescu, Basarab. 2002. *Manifesto of Transdisciplinarity*. Translated by Karen-Claire Voss. Buffalo: SUNY Press.

Northoff, Georg. 2004. "Humans, Brains, and Their Environment: Marriage between Neuroscience and Anthropology?" *Neuron* 65, no. 6: 748–51. doi:10.1016/j.neuron.2010.02.024.

Persinger, Robert. 2001. "The Neuropsychiatry of Paranormal Experience." *Journal of Neuropsychiatry and Clinical Neurosciences* 13 (November): 515–24.

Restek, Richard. 2006. *The Naked Brain: How the Emerging Neurosociety Is Changing the Way We Live, Work, and Love.* New York: Three Rivers Press.

Richardson, Alan. 1999. "Cognitive Science and the Future of Literary Studies." *Philosophy and Literature* 23, no. 1: 157–73.

Roepsdorff, Andreas. 2007. "The Neuroturn: Challenging Anthropology or Anthropological Challenge?" Lecture presented at the Lecture Series on Culture and Cognition, London School of Economics, January 17.

Rorty, Richard, ed. 1967. *The Linguistic Turn: Essays in Philosophical Method.* Chicago: University of Chicago Press.

Roskies, Adeline. 2002. "Neuroethics for the New Millenium." *Neuron* 35, no. 1: 21–23.

Salisbury, Laura, and Andrew Shail, eds. 2010. *Neurology and Modernity: A Cultural History of Nervous Systems, 1800–1950.* Houndsmills: Palgrave.

Sample, Ian. 2007. "The Brain Scan That Can Read People's Intentions." *The Guardian* February 9. Available at http://www.guardian.co.uk/science/2007/feb/09/neuroscience.ethicsofscience.

Saunders, Barry. 2008. *CT Suite: The Work of Diagnosis in the Age of Noninvasive Cutting.* Durham: Duke University Press.

Schwartz, Jeffrey M., and Sharon Begley. 2003. *The Mind and the Brain: Neuroplasticity and the Power of Mental Force.* New York: Harper Perennial.

Schwartz, Jeffrey M., and Beverly Beyette. 1997. *Brain Lock: Free Yourself from Obsessive-Compulsive Behavior.* New York: Harper Perennial.

Stokols, D., K. L. Hall, B. K. Taylor, and R. P. Masert. 2008. "The Science of Team Science: Overview of the Field and Introduction to the Supplement." *American Journal of Preventative Medicine* 35, no. 2S: S77–89.

University of Pennsylvania. 2011. "Neuroscience Boot Camp" Available at http://www.neuroethics.upenn.edu/index.php/events/neuroscience-bootcamp.

Vidal, Fernando. 2009. "Brainhood, Anthropological Figure of Modernity." *History of the Human Sciences* 22, no. 1: 5–36.

Wilson, Robert A. 2004. "What Computations (Still, Still) Can't Do: Jerry Fodor on Computation and Modularity." *Canadian Journal of Philosophy* 30:407–35.

Zeki, Semir. 1998. "Art and the Brain." *Daedalus* 127, no. 2: 71–104.

Part I

The Neuroscientific Turn in Context

When you have a phenomenon that inspires phrases like "the neuro-revolution" and, yes, even "the neuroscientific turn," it is hard not to rush to classify all things neuro- as novelties at best, fads at worst—the brightest area of inquiry with a shiny assortment of toys, techniques, and theories. Yet, as Fernando Vidal points out (2009), the hype emerging from the Decade of the Brain is but an echo of earlier claims that the brain ought to move to the forefront of biomedical research. Thinking historically offers us a way to "evaluate the contextual contingencies of the experimental concepts, clinical categories and methodologies, and consider new ways to study, apply or interpret them," and it can also "act as a counter-force to the 'hype' surrounding many themes in the neurosciences and contribute to more cautious portrayals and realistic expectations" (Choudhury, Nagel, and Slaby 2009, 66). The essays in this section should not be confused with a history of neuroscience per se; instead, they should be understood as a selected genealogy of earlier and alternative intersections between the brain sciences and other discourses: nineteenth-century literature (Murison); sixteenth-century dissection, typography, and faculty psychology (Kismet Bell); and, more recently, post-poststructuralist philosophy (Gotman). Each essay reminds us that our resurgent interest in neuroscience is not radically new, not a seismic shift, but a series of aftershocks.

Chapter 1

"The Paradise of Non-Experts"

The Neuroscientific Turn of the 1840s United States

Justine S. Murison

To examine a neuroscientific "turn," as this collection aims to do, is to give a particular shape to intellectual history. As opposed to a "revolution," a "turn" signifies a quieter if no less profound pivot. Pivots turn away from and turn toward, in this case away from psychoanalytic models and toward contemporary neuroscientific ones. Current fascination among humanists and social scientists with neuroscience arguably came at a moment of exhaustion with a century-long attachment to Freudian psychoanalysis. Long a rich language of the psyche as expressed in somatic traces, psychoanalytic terms like *anxiety* and the *unconscious* have persisted in the theoretical language of the humanities and social sciences even as strict psychoanalytic readings fell out of favor (and certainly long after its usefulness in therapeutics had largely waned). Neuroscience refreshes the methodological assumptions of humanists and social scientists by providing both a vocabulary and a model for reconsidering the relation of psyche and soma. It encourages scholars to consider the very *materiality* of the brain and nervous system as the location of complexity and creativity rather than the screen of a "deeper" psyche, which is the predominant psychoanalytic model (Massumi 2002; Wilson 2004; Connolly 2002; Sedgwick 2003). Neuroscience differs from psychoanalysis and strictly cognitive science by emphasizing the brain as plastic and the neurology of the body as preceding and forming rationality (Damasio 1999). In turn, the humanities (and their reach into popular culture) operate not simply to "popularize" the discoveries about the brain in the last two decades; rather, they constitute the modes by which a new language of the self, based in neuroscience and its therapeutic corollaries like pharmacology, invest everyday people with an understanding of their own brains and emotions as manageable material to be transformed.

Focusing on a historical precedent for this recent neuroscientific turn, this essay explores the dynamic cultural conversation that a preoccupation with the nervous system can produce. The United States of the 1840s is a crucial case study for the current interest in neuroscience, for it too experienced an unprecedented explosion of interest in the workings and meanings of the nervous system. The widespread popularity of neurology in the 1840s owed much to the men and women who publicly performed mesmerism, later renamed hypnosis, for eager antebellum audiences. Popular demonstrations of trance predated modern disciplinarity and can thus represent a historical moment when attention to the nervous system came from a multitude of directions and impacted not just science but also literature, politics, and belief. More often than not, mesmerists focused on the capacities of the nervous system to cohere groups, to provide access to others' minds, and to glimpse spiritual realms. Mesmerism was both a complex science of the nerves and a popular entertainment, and thus its role in the rise of a nineteenth-century language of the nerves has been overlooked (and too easily relegated to the category of "pseudoscience"). Most important, the professionalization of American neurology in the late nineteenth century by neurologists such as George Miller Beard and S. Weir Mitchell depended upon an opposition between scientific studies of the nervous system and mesmerism's popular associations.

I make two arguments in this essay: first, that nineteenth-century mesmerism acted as a conduit between medical circles and the public, helping to make knowledge about the physiology of the nervous system widely available; and second, that in popularizing this new version of the self—rooted in nervous physiology—advocates of mesmerism promoted a "materialized spirituality," in which the materials of the body could replace the immaterial soul as the source of spiritual wonder and faith. In turn, this fusion of the spiritual and material exemplified by mesmerism played a significant though often tacit role in the professionalization of neurology, for in blocking off the spiritual aspects mesmerism represented, neurologists in the later part of the century consolidated their professional authority. Finally, I end this essay with a return to the issue of recent interdisciplinary work that echoes this earlier, predisciplinary era, to comment upon both the promise of this moment and its significant limitations and differences from its 1840s precedent.

Mesmerism, animal magnetism, hypnotism, all three of these nineteenth-century terms refer to the process by which one person (often called the mesmerist or magnetist) put another person (variously termed the subject, the somnambule, or the patient) into a suggestive sleep that then may give the mesmerist influence over the body of that person. Whether on the stage or in the sickroom, mesmerism would typically begin with what were called "passes," the movement of the mesmerist's hand or an object to induce trance. By probing such subjects as group psychology or providing much-needed relief for those suffering from a variety of incurable and painful illnesses, from migraines and neuralgia to cancer, mesmeric demonstrations claimed to be both philosophical and medical. In the United States, the height of mesmerism as both a complex science and a mass entertainment occurred in the 1840s. Mesmerism flourished in Jacksonian America, roughly the period from Andrew Jackson's 1828 election through the 1840s, precisely because this era was the nadir of medical professionalism in the United States. In an era marked by profound distrust of elite institutions and professional expertise, medical licensure receded. Those few states with licensing laws, in fact, rescinded them, and medical licensure was not to become standard until the 1870s (Baker 1984; Conrad and Schneider 2009; Duffy 1993; Starr 1982).[1] Thus the antebellum medical community could not cordon off its profession from challenges and encroachments from "irregular" practices like mesmerism. Not coincidentally, then, the rise of mesmerism occurred at a time when more citizens participated in scientific culture than they would by the turn of the twentieth century. The relegation of mesmerism at the end of the century into merely a laughable and possibly fraudulent pastime, in turn, coincided with the professionalization of medicine, and in particular the professionalization of neurology as a unified discipline.

The history of mesmerism reaches back to Paris at the end of the ancien régime. In his medical dissertation, Viennese physician Franz Anton Mesmer theorized the presence of "animal gravity" within people's bodies, which he later developed into a fuller medical theory and therapeutic regimen (Albanese 2007; Darnton 1968; Fuller 1982; Gauld 1992). According to Mesmer, "animal gravity" consisted of a magnetic fluid that passes from the magnetist to the patient, and can also pass between people, much as electricity does (see fig. 1). He moved to Paris in 1778 and became a prominent figure there, treating patients and publicly demonstrating his theory. These exhibitions were flamboyant, to say the least. They included a "daisy chain" of people all mesmerized with the same application of the

Fig. 1. "Animal Magnetism—The Operator Putting his Patient into a Crisis." Engraving by Daniel Dodd, in *A Key to Physic, and the Occult Sciences,* by Ebenezer Sibly (1802). (Image courtesy of the Wellcome Library, London.)

"fluid" and the (easily lampooned) mesmerization of trees (Darnton 1968, 8). In 1784, a Royal Commission composed of some of the most prominent men at court—including Benjamin Franklin and Antoine Lavoisier—investigated Mesmer's claims. They experimented with his theories of trance, listened to his lectures, and finally declared him a fraud, and the trance effects to be the result of overheated imaginations. Yet even as the commission attempted to have the final word, mesmerism never entirely disappeared either on the Continent or in England.

Though mesmerism was a great source of European speculation and entertainment during the revolutionary era, it failed to catch the attention of the newly formed United States. By the 1830s, though, the United States was ready for mesmerism. This was the era of Jacksonian democracy, which promoted the "common people" over the elite, and of transcendentalists such as Ralph Waldo Emerson, who urged his fellow citizens to experience nature through immediate contact. Mesmerism joined these ideologies as a democratic science. Following the itinerant lectures of French mesmerist Charles Poyen in the Northeast during the 1830s, both physicians and lay practitioners began to practice and debate mesmerism (Fuller 1982; Gauld 1992; Crabtree 1993; Albanese 2007). By then, mesmerism had far surpassed the theories of Anton Mesmer. The new mesmerists of these decades depended upon the discoveries of nervous physiology to explain the phenomenon of mesmeric sleep.

By 1840, nervous physiology was an exciting but contentious field of study. Unable to discern how the nerves operated, physicians did believe that some sort of fluid, akin to electricity and maybe electricity itself, fueled the system. "Animal magnetism" thus operated as both a phrase to describe the function of the nervous system in its most mysterious form and also a term for the supercharged dynamics of mesmerism, the ability of one person's magnetic or electrical qualities to influence another. Thus electrical metaphors traveled well beyond either physicians or mesmerists, and in the case of electrical therapeutics—much milder than their twentieth-century corollaries—they exceeded metaphor to become literal (Otis 2001; Gilmore 2009). Just like regular physicians, mesmerists depended upon metaphors of electricity and electrical technologies such as the telegraph to explain the operations of the nerves. For instance, John Bovee Dods, a former Universalist minister and self-proclaimed "electrical psychologist," lectured on the wonders of mesmerism and animal electricity to the U.S. Senate in 1850. In this series of lectures, solicited by Henry Clay and Daniel Webster (among others), Dods argued that the "brain was invested with a living spirit" and "like an enthroned deity, presides over, and governs, through electricity as its agent, all the voluntary motions of this little, organized, corporeal universe" of the human body (Dods 1850, 148).

Dods's lectures at the U.S. capital should be some indication of the widespread reach of mesmerism by 1850. Mesmerism was both a scientific pursuit and an occasion for the gathering of all types of people in both public and private spaces for the inspection of the powers of the body. Mesmerism flourished through stage exhibitions like those of Professor Rodgers and Mrs. Loomis (see fig. 2), but, more significantly, its reach

was due to the print world of antebellum America, including most pre-
dominantly the extensive circulation of magazines and newspapers. A wide
range of unlikely periodicals published vibrant discussions on mesmeric
phenomena, including the *Christian Examiner, Godey's Magazine and Lady's
Book*, the *Maine Farmers and Mechanics Advocate*, and William Lloyd Garrison's *The Liberator*, to name only a few. The popular understanding of
the nervous system derived from mesmerism's extensive permeation of
print culture. A limited study of the period between January 1, 1844, and
December 31, 1845, in which Edgar Allan Poe published three of his most
popular mesmerism tales—"The Facts in the Case of M. Valdemar," "Mesmeric Revelations," and "A Tale of the Ragged Mountains"—in the American Periodical Series Online database turned up 200 articles, a number
that includes original works and reprints from England and Europe.[2] If we
count circulated reprints (i.e., repeated publications of the same article or
story) that had their original publication date in 1844 or 1845, the number
stretches to 253.[3] These citations included medical reports about mesmeric
surgeries, polemics against mesmerism, squibs reporting mesmeric exhibits
and publications, book reviews, plays, poems, and fiction.

In 1844, the first reprint in the United States of British writer Harriet Martineau's defense of mesmerism, published originally in the London
Athenaeum that same year, appeared in *The Phalanx: Organ of the Doctrine
of Association*, a tiny journal promoting the French utopian theory of Fou-
rierism. From there, Martineau's letter on mesmerism would be reprinted
at least twenty times in this two-year span, often with extensive commentary by editors, and this number does not count essays and editorials in
response to her argument. Second only to Martineau's presence in these
two years was Edgar Allan Poe, whose "Mesmeric Revelations" and "The
Facts in the Case of M. Valdemar" recounted tales of mesmerized speech
from beyond the grave. "Valdemar" was even reprinted in pamphlet form
in England with the title changed to "Mesmerism in Articulo Mortis: An
Astounding and Horrifying Narrative," suggesting, in essence, that the
story was one of fact, not fiction.

This quantification of the archive merely scratches the surface. Mesmerism was a deeply contentious science both within its own circle and
at the boundary it shared with medicine. The debates about mesmerism
concentrated on how it worked or whether it was a fraud, and these debates
were conducted across different journals and even, on occasion, within
them. Sometimes opposing strands of the debate appeared in the same
issue of a journal. For instance, the December 28, 1844, issue of the *Medical
Examiner and Record of Medical Science* included two articles within pages of

GREAT EXCITEMENT!
PHRENOLOGY
And Animal Magnetism!

At the *particular request* of the *audience*, last *Wednesday evening*.

MRS. LOOMIS,
WILL LECTURE UPON
PHRENOLOGY,
THIS SATURDAY EVENING,
AT THE CHINESE MUSEUM LECTURE ROOM,
ENTRANCE IN GEORGE STREET,

Subject—the organs of the brain. After which she will examine Phrenologically one of the audience, who will report
whether the examination is correct.

PROF. RODGERS,

Will then perform many new and interesting experiments in

ANIMAL MAGNETISM,

with MASTER MILLER and MR. E. HORN and conclude
by magnetising the audience. Commencing at 7½ o'clock.

Tickets twenty-five cents, or five for one dollar.

Several persons will be thrown into the

SPIRITUAL STATE,

exhibiting many curious Phenomena, claiming the
attention of all scientific persons.

The experiments will exhibit: The Concentration of the Nervous force, Muscular action, Nervous congestion, Important application of this
Law in Disease, Sympathy, Antipathy, Laws of
Magnetism, Art of Magnetising explained, great
majority subject to the Magnetic influence, the
Magnetic chain of recipients,

DELUSION, SUBLIMITY, TRANCE, CLAIRVOYANCE,

Phreno-Magnetism, Transposition of the Senses,
MAGNETISED WATER,

MAGNETISING AT A DISTANCE,

Experiment with the Handerchief, Paralysis,
Examining the sick, Insensibility to pain,
Vital electricity, Neurology, &c., &c.

At the conclusion PROF. RODGERS will magnetise a number
of the audience as they sit, showing that this has been performed
by him some time previous to the attempt of REV. LE ROY
SUNDERLAND, at Lowell, by what he denominates Pathetism.

**Doors open at half past 6 o'clock, Lecture will
commence at half past 7 o'clock precisely.**

Tickets for sale at COLON'S, 203½ Chesnut Street, Professor RODGERS' Rooms, Union
Buildings, Eighth and Chesnut Street, Third floor, and at the Lecture Room, George
Street, on the night of the Lecture.

BROWN, BICKING & GUILBERT, PRINTERS, NO. 56 NORTH THIRD STREET, ABOVE ARCH, PHILADELPHIA.

Fig. 2. "Great Excitement!
Phrenology and Animal
Magnetism!" Broadside,
Philadelphia, 1843. (Image
courtesy of The Library
Company of Philadelphia.)

each other: F. S. Burman's "More Mesmeric Imposters" and an anonymous report titled "Triumphs of Mesmerism."

For the most part, though, physicians sounded alarms about fraud. A contributor to the *Boston Medical and Surgical Journal* raged, "It is the ignorant portion of the community, many of whom earn their bread by the sweat of the brow—an honorable course of life—who are almost exclusively taxed for the support of those travelling, vagabond mountebanks, who call themselves professors of Mesmerism" (*Boston Medical and Surgical Journal* 1842, 110). David Meredith Reese, a New York physician, snidely declared that mesmerists were simply "third rate doctors, merchants and mechanics" who have failed in their original pursuits and supplied themselves with a "factory girl, who would *rather sleep than work*," in *Humbugs of New York* (1838, 35).[4] Without the clout that licensure would give physicians later in the century, though, doctors like Reese had no way to exclude mesmerism from the circle of "sciences." Other physicians were less explosive and sought to agree with mesmerists that they were experimenting on the nervous system. This was physician Reynell Coates's argument in his lecture to the Philadelphia Medical Society in 1841. These medical mysteries, according to Coates, might be "illuminated by the prosecution of truly philosophical investigations into the history of this mutual nervous influence, to which it is at present an utter folly to attach the names of electricity, magnetism, galvanism, or gravitation, there being at present not the shadow of a reason for such connexion" ("Proceedings" 1841, 794). According to Coates, medical experimentation ought to solve the mysteries of the nervous system and consolidate the various unregulated terms and theories into one medical explanation.

Many antebellum citizens shared Reynell Coates's optimism that medicine was on the cusp of explaining the mysteries of the body by integrating into itself these popular sciences, but this medical idealism would not survive the Civil War. In fact, one of the most prominent American neurologists of the postbellum era, George Miller Beard, would study mesmerism, trance, mind reading, and spiritualism only to cordon off their findings from his own in his most famous contributions to late nineteenth-century neuroscience, *A Practical Treatise on Nervous Exhaustion (Neurasthenia): Its Symptoms, Nature, Sequences, Treatment* (1880) and *American Nervousness: Its Causes and Consequences* (1881). These two texts together argued that Americans (both men and women) suffered from a pathology called "neurasthenia," which Beard defined as a "*lack of nerve-force*," a disease occasioned by the modernization of American life (Beard 1972, 5). He argued that the fatigues of modern life—from political tumults to technological innovations—produced nervous citizens. This description of the United States as a nation prone to nervousness was the least innovative aspect of Beard's argument. He simply consolidated the general nineteenth-century medical view of nervousness under the neologism *neurasthenia* (Rosenberg 1997). Beard's contribution to neurology, though, was remarkable for its specific rejection of mesmerism (and its fellow travelers like spiritualism and mind reading) and that practice's cultural ubiquity. In fact, what separated Beard from his antebellum predecessors was how he distanced himself from what he considered to be the democratic excesses of their empirical methods.

According to Beard, solving the mystery behind "parlor tricks" like mesmeric trance and mind reading was "one of the most important problems of science" on account of its "relations to physiology and pathology" and "its bearings on the principles of evidence, the estimate of human testimony" (1877, 1–2). As Beard put it, "If trance, the involuntary life, and human testimony, were understood universally as they are now beginning to be understood by students of the nervous system, there would not, could not be a spiritist on our planet; for all would know that spirits only dwell in the cerebral cells—that not our houses but our brains are haunted" (1879, 67). In particular, he targeted the emphasis on empirical evidence in the popular sciences. "So far as the senses are concerned," Beard pronounced, "they deceive all of us every hour and every moment" and therefore "seeing is not believing, but doubting; for what is all human science but a correcting of the errors, and a supplementing of the defects of the senses" (1877, 4). Beard ultimately rejected "mere" sensual empiricism and thus the antebellum excitement over the spiritual and moral possibilities of popular forms of neuroscience (Taves 1999; Brown 1983; Shortt 1984). Calling the antebellum period "the paradise of non-experts," Beard lambasted the era for confusing "a mass of empiricism—a chaos, dark, formless, boundless, inaccessible to science" with science (and specifically neurology) itself (1879, 72, 70). Beard subsequently predicated *American Nervousness* on the assumption that *induction* and logical reasoning must always frame discussions of the nervous system, rejecting the overly confident reliance on human testimony and physical evidence that flirted with popular stage entertainments of the first half of the nineteenth century. *American Nervousness* was, above all, a cultural explanation of neurasthenia that promoted medical expertise over and above the knowledge of patients and irregular medical practitioners.

Beard's attempt to separate his version of neurology from the antebellum "mass of empiricism" testified to a moment of disciplinary consolidation around the study of the nervous system in the late nineteenth century. The New York Neurological Society was founded in 1872, the *Journal of Nervous and Mental Disease* began publishing in 1874, and the American Neurological Association had its first meeting in 1875 (Bailey 1951; DeJong 1982). Beard joined the American Neurological Association the year it was founded and thus participated in this early consolidation of neurology as a coherent discipline out of what had once been a variety of pursuits. In the antebellum period, in contradistinction, the various studies of the nervous system reflected what Thomas Kuhn would term a "pre-paradigm period," in which no one paradigm has achieved validity and thus different participants—including those within and outside of medicine and science—could join the field of study (1962). Indeed, antebellum mesmerists certainly considered themselves to be scientists and, in many cases, physicians. They thus actively debated the causes and advanced their own terms for the physiological effects of trance. While James Stanley Grimes suggested his fellow mesmerists adopt "nervous fluid" as the best term for the agent of the nervous system, John Bovee Dods persisted in promoting, as we have seen, "electrical psychology." Famed mesmerist La Roy Sunderland (known in the antebellum period for his purported ability to put whole audiences into a trance) dismissed these terms and advocated his own awkward one, "pathetism," which he described as a type of emotional and imaginative sympathy between people. While none of these practitioners "won" the ability to claim their vocabulary triumphant in the long run, this debate crossed into medicine, as regular physicians also discussed whether "animal electricity" or the "nervous fluid" fueled the nervous system and how trance was made possible by the physiology of the body. Debates between mesmerists over what caused trance and how best to label that physiological impulse dissolved in Beard's work into a strict line between expert and amateur. Denying the voice of patients and sundry amateurs like poets, ministers, or herbalists, Beard sought to distance his work from naive empiricism.

Beard's rejection of antebellum science thus had everything to do with that era's skeptical view of expert knowledge and its simultaneous (and paradoxical) embrace of scientific method. Perhaps no event in 1840s America better exemplified the era's predisciplinary and antiexpert environment than the arrival of Joseph Rodes Buchanan in New York City. A Kentucky physician, Buchanan spent the early years of his career formulating a theory of "neurology." Buchanan's neurology sought to demonstrate phrenology (reading the bumps of the head for insight into character) through mesmerism by, as he suggestively put it, "inspiring the organs *to speak for them-*

selves" through mesmerizing discrete portions of the brain (1842, 13). As he explained, "The term 'Neurology,' by signifying the science of the nervous system, is competent to embrace all its functions, as well the mental as the corporeal, and is therefore the proper term for that comprehensive science, of which Craniology, Phrenology, and Physiology, are constituent portions" (23–24). Buchanan's theory of neurology gathered together the disparate pursuits of mesmerism, animal magnetism, and phrenology to forge one discipline of neurology. Yet he did so from a perspective at which Beard would later scoff: promoting a democratic availability, from the Eclectic Medical Institute he helped found in Ohio to the democratic rhetoric of allowing the organs to, as he put it, *"speak for themselves."*

This vision of democratic voices guided the New York showcase of Buchanan's theory of neurology. On November 4 and 5, 1843, a publicly selected committee met to witness and judge Buchanan's experiments in neurology. We would now refer to the subcommittee as "interdisciplinary": William Cullen Bryant, one of the decade's most prominent poets and editor of the *New York Evening Post;* John O'Sullivan, editor of the *Democratic Review* and coiner of the term *manifest destiny;* Dr. Samuel Forry, widely read in this decade for his studies on climate; and Henry Whitney Bellows, Unitarian minister who would go on to establish the United States Sanitary Commission during the Civil War. Bryant and O'Sullivan published the committee's report in their respective newspaper and journal. The report found "neurology" promising and the experiments on several patients convincing. Widely read, the report even made fictional appearances. In a satiric short story, "The Mysterious Neighbor," first published in the *Columbian Lady's and Gentleman's Magazine,* a main character announces his intention to introduce "a new theory, or science, which I mean to bring out some of these days, to compete for public favor and the approbation of Messrs. Bryant and O'Sullivan, with animal magnetism and neurology" (Pencil 1844, 201). The satire here covertly registers what the Buchanan report manifests: that in an era so hostile to expertise in general, science depended upon a wider audience for legitimacy.

Parodies about the culture of mesmerism like this moment in "The Mysterious Neighbor" or Poe's "The Facts in the Case of M. Valdemar" (1845) abounded. These parodies could be so subtle that audiences mistook them for fact. For example, in Poe's tale, Valdemar agrees to be mesmerized while dying, and his body hovers between life and death in a mesmeric trance for nearly seven months before rotting away into putrefaction. Despite the fact that "Valdemar" would seem wholly unbelievable to an audience today, it convinced several of Poe's contemporaries. One of those to take it seriously was English mesmerist Robert Collyer, who wrote to the *Broadway Journal* (the magazine that originally published Poe's story and which he edited) to explain that he has not the "least doubt of the *possibility* of such a phenomenon; for, I did actually restore to active animation a person who died from excessive drinking of ardent spirits" (1845, 390). Likewise, Poe's future fiancée Sarah Helen Whitman begged a mutual friend to find out whether Poe's "Mesmeric Revelations," his other popular mesmeric tale, was based on fact (Silverman 1991, 348).

Although Helen Whitman's query may be easily dismissed today, she knew that stories like "Valdemar" and "Mesmeric Revelations" asked pressing philosophical and religious questions. Indeed, fictional tales of mesmeric demonstrations did more than testify to the extent of the cultural excitement over this science and the satire that excitement could inspire. In the hands of Poe, they also drew together very serious concerns of faith, evidence, and professional authority that the popular science of the nerves raised. In this story, which reached a wide reading audience in Britain and the United States, another Dr. P (we can assume Poe is using "P" with tongue in cheek) converses with Mr. Vankirk, whom he has mesmerized in his death throes. From this intermediary state, Vankirk voices many of Poe's seemingly more earnest beliefs about God. God, Vankirk explains, "is not spirit, for he exists. Nor is he matter, *as you understand it.* But there are *gradations* of matter of which man knows nothing; the grosser impelling the finer, the finer pervading the grosser" (Poe 1984b, 720). In this scale of matter, God is the "ultimate, or unparticled matter," which "not only permeates all things but impels all things" (720). With Poe, one never knows when he is parodying (as James Russell Lowell once put it, Poe was "three fifths . . . genius and two fifths sheer fudge"), but this theory of gradations of matter into spirit was one of the bases of his prose-poem *Eureka*. Clocking in at over a hundred pages, *Eureka* begins with satire but quickly shifts tone to a philosophical *tour de force* about the universe and God. Poe's descriptions of God as spiritual matter in both "Mesmeric Revelations" and *Eureka* harmonize perfectly with the most prominent descriptions of the "nervous fluid" as a material invisible to the eye but apparent in the body. Matter, by 1848 when Poe wrote *Eureka*, can be both spiritual and material.

If mesmerism was how antebellum Americans learned to speak about the nervous system, as I have been arguing, it led to more profound questions about the relation of mind to matter and empiricism to faith, as Poe's "Mesmeric Revelations" suggests. Indeed, belief was the pressing question for many antebellum Americans as they visited mesmeric demonstrations or read about them in the era's periodicals. In the 1840s, prominent author and abolitionist Lydia Maria Child turned directly to faith in response to mesmerism. "There is something exceedingly arrogant and short-sighted in the pretensions of those who ridicule everything not capable of being proved to the senses," Child announced in a series of essays on New York City for the *Anti-Slavery Standard*, the newspaper that she edited from 1840 to 1843 (Child 1998, 83). In this particular letter, she urged skeptics to "be humble enough to acknowledge that God governs the universe by many laws incomprehensible to you; and be wise enough to conclude that these phenomena are not deviations from the divine order of things, but occasional manifestations of principles always at work in the great scale of being, made visible at times, by causes as yet unrevealed" (Child 1998, 85). Child's defense of the unseen was not pitched as a battle between religious faith and science; rather, it was a response to medical skepticism about animal magnetism. For Child, the question was not *whether* she believed in mesmerism but what must change methodologically in science for others to understand how it occurs.

The dismissive attitude Child bemoans was what Harriet Martineau confronted in *Life in the Sick-Room* (1845), a collection of essays that included her noteworthy defense of mesmerism circulating in the United States in 1844. She proclaimed, "Whatever quackery and imposture may be connected with" mesmerism, "however its pretensions may be falsified, it seems impossible but that some new insight must be obtained by its means, into the powers of our mysterious frame" (1845, 83). Martineau turned to mesmeric passes for relief from a prolapsed uterus caused by an ovarian cyst. Unable to find a mesmerist to attend her, Martineau and her maid succeeded in imitating mesmeric passes on their own, which relieved her pain. The urgency of Martineau's commentary came from her position as a patient disenchanted with regular medicine. She offered her testimony, she explained, on the "ground that though the science of medicine may be exhausted in any particular case, it does not follow that curative means are exhausted" (1844, 323). Her target in this essay was what she called the "easy mockery" of patients in pain by those who are healthy or by medical experts more concerned with their own professional status than providing patients with relief. Martineau juxtaposed what she saw as mesmerism's pragmatic and palpable results as she experienced them to the abstracted pursuit of facts apart from the needs of patients.

Referring to the penchant of physicians to dismiss mesmerism (among other medical "fads" such as phrenology and the water cure), Martineau asked, "[Are] we not learning, from their jumbled discoveries and failures, that empiricism itself is a social function, indispensable, made so by

God, however ready we may be to bestow our cheap laughter upon it?" (1845, 84). Martineau celebrated empiricism as a "social function"—one that depended upon all members of society—and thus exemplified Beard's reviled "non-experts." This active role, she believed, was imperative to the development of science. She ended her essay reminding others who have found relief from their complaints through the aid of mesmerism that "near them there exist hearts susceptible of simple faith in the unexplored powers of nature, and minds capable of an ingenuous recognition of plain facts, though they be new, and must wait for a theoretical solution" (1844, 323). Martineau's final thoughts typify the methodological struggle over mesmerism, which was not just about unease with medical professionalism but about the impact of new theories of nervous physiology on questions of both fact and faith. One of the effects of the popular debates on mesmerism therefore was a realignment of medical rationalism and spiritual faith.

Child's and Martineau's claims that mesmerism can be a way to maintain faith within a nineteenth-century scientific culture seem paradoxical at first blush. For many observers, mesmerism (like studies of the nervous system in regular medicine) turned the mind and soul into material rather than spiritual elements, which, in turn, risked the basic premise of the persistence of the soul in the afterlife. For instance, Nathaniel Hawthorne famously warned his fiancée Sophia Peabody of this danger. When Sophia sought relief from her migraines through the treatments of her close friend and newly established mesmeric healer Cornelia Park, Hawthorne wrote her not to "let an earthly effluence from Mrs. Park's corporeal system bewilder thee, and perhaps contaminate something spiritual and sacred. I should as soon think of seeking revelations of the future state in the rottenness of the grave" (Hawthorne 2002, 96). But for Poe, Martineau, and Child, mesmerism—and the nervous system it demonstrated—afforded a glimpse of something beyond the material even though grounded in the materials of the body. It therefore positioned the body, and not the soul, as the spiritual element sought in both philosophy and religion. For all three, mesmerism challenged the primacy of medical materialism from within that discourse; sharing the medical assumptions about the nervous system but concluding that the nervous system revealed aspects of ethics, faith, and the afterlife, these writers and reformers shaped a popular, spiritual conception of neuroscience as based on human experience and as democratically available to all.

Considering how lively the debate over mesmerism was in 1840s America, we might call the recent "neuroscientific turn" a *re*turn to an era before medical institutions cordoned off the discoveries about and treatments of the nervous system to a distinct discipline engaged in and monitored by professional scientists and physicians. As with any return, however, this is a return with a difference, perhaps in instructive ways. Unlike in the 1840s, our most recent turn to neuroscience comes with the medical and scientific legitimacy accrued over the twentieth century (Conrad 2007; Conrad and Schneider 2009; Freidson 1970). According to Peter Conrad, the twentieth century was the era of medicalization, in which "nonmedical problems become defined and treated as medical problems," a process made possible by the increased intrusion of medicine into all aspects of everyday life (2007, 4). More particularly, the turn to neuroscience in the humanities and social sciences arises out of a shift from medicalization to "biomedicalization," which is a way to describe the pervasive power of technoscientific innovations over common life that leads "health itself and the proper management of chronic illnesses" to become "individual moral responsibilities to be fulfilled through improved access to knowledge, self-surveillance, prevention, risk assessment, the treatment of risk, and consumption of appropriate self-help/biomedical goods and services" (Clarke et al. 2003, 162). Both medicalization and biomedicalization can be seen as integral aspects of what Michel Foucault called "biopower," that is, the state's investment in the control of populations through biological technology. Where neuroscience meets biomedicalization, in particular, is in how it monitors and judges not just disorders but healthy systems. Thus by chronicling and controlling healthy nerves neuroscience offers the promise of self-help through brain imaging and other technological achievements.

The major distinction between the height of mesmerism and current excitement over neuroscience is what allowed for the legitimacy of medicalization and biomedicalization: the rise of scientific medicine in the mid-nineteenth century (Rosenberg 1992). There is no doubt that any nostalgia for the democratic possibilities of science in the 1840s must be tempered by the recognition of such developments as the germ theory of disease and other crucial insights made possible by treating medicine more as a science than an art. The rise of scientific medicine was aided by technological innovation, from microscopes to computerized axial tomography (CAT) scans. Perhaps the most dramatic difference between these two "neuroscientific turns" is, in fact, located in technology. Whereas it was feasible for a minister or essayist (or even Martineau's maid) to become a practical mesmerist, the barriers for the participation of humanists and social scientists in neuroscience are now quite profound. MRI technology, to give one instance, is prohibitively expensive, even for many scientists, physicians, and hospitals.

A machine can cost between one and three million dollars, and this price does not include the necessary concrete enclosures or the funds needed to run and maintain it. In contrast, mesmerism barely cost anything, and the demonstrations, on the other hand, could well reap funds. As much as "interdisciplinary" work is touted in the modern university, there is no return to an era of widespread and popular scientific experimentation such as Americans experienced in the 1840s.

The other major distinction between these two neuroscientific eras relates to the cultural appreciation of science. The new fascination with neuroscience redoubles scientific legitimacy through popular appeals to evidence provided by MRI technology and pharmacology. Yet it may be instructive to recall the paradox of the 1840s, when scientific method and nervous terminology rose in popular opinion, but professional medicine declined in legitimacy. Far from a uniform biomedicalization, the modes of popular appropriation of these sciences can far exceed their disciplinary origins or aims, just as mesmeric trances opened the door for both spiritual imagining and critiques of regular medicine. In other words, thorough biomedicalization may well be illusory, and humanists, especially, may be well trained to see how the new discoveries in neuroscience follow unpredictable paths into popular culture, offering a model of the mind-body nexus that is rooted in science but helps the culture reconsider its relation to science.

These distinctions between the 1840s and today should not then overshadow what might be the most important connection. In the case of neurology, what is most strikingly similar between then and now is the excitement and joy that humanists and social scientists register in this reinvigorated model of the brain. As political scientist William Connolly puts it, he seeks in neuroscience "sustenance" with which to "reflect upon perception, thinking and time in politics" (2002, 2). Connolly is more than ready to point out that he is no neuroscientist and that his project is not identical to that of neuroscience, yet he uses neuroscience to reimagine the questions of traditional political science. Brian Massumi turns his use of neuroscience to account in a different way, by suggesting that "poaching" a scientific concept will help make the humanities "differ from the sciences in ways they are unaccustomed to," a process that may help nonscientific disciplines "rearticulate what is unique to their own capacities" (2002, 21). The antebellum United States may well model a salutary way in which "nonexperts" (like today's humanists and social scientists) can engage neuroscience to such effects. For Poe, Child, Martineau, and even the splenetic George Beard, the questions they asked were about methodology itself:

what can we learn from our senses and what is the relationship between everyday experience, physiological discovery, and metaphysical questions of health, belief, and ethics? Poe's Dr. P sparked conversations about the ethics of experimentation; Martineau urged physicians to pay more attention to the voices and experiences of patients; and Child reminded medical skeptics that the nervous system is a powerful example of the limits of knowledge, as the brain and nerves still remain today. Perhaps most of all, mesmerism can return contemporary neuroscience to its hidden but crucial historical trajectory, a history fundamentally inflected by the boundary concerns of disciplines. Thus, the most recent "neuroscientific turn," the subject of this collection, is not so much a revolution in disciplinary methodology but the latest iteration of a long-standing exchange between the humanities, social sciences, and natural sciences on the role of the body and brain in culture and politics.

NOTES

1. The American Medical Association (AMA) formed in 1847 as an explicit response to "irregular" medical practices of the era. In the United States, it would not be until the 1870s and 1880s, though, that medical professionalization achieved its aims, in particular the creation of licensure laws that would transform "irregular" medicine into "unlicensed" (and thus illegal) medical practice. Neurologists benefited from this larger movement, which was led by bacteriology, surgery, and the general rise of scientific medicine in the wake of germ theory. See Conrad and Schneider 2009, particularly 194–98.

2. The American Periodicals Series Online is a digital database (originally a microform collection) of over 1,500 magazines and periodicals from 1740 to 1900. The first half of the nineteenth century includes 900 periodicals. My methodology consisted of keyword searches on variations of "mesmerism" and "animal magnetism," and a double-checking for reprints. All of the articles counted directly address mesmerism or its synonym "animal magnetism." I did not count stray uses of the terms. I have also excluded advertisements, tables of contents, indexes, and back material.

3. This is the era of extensive reprinting, as Meredith McGill has shown. Reprinting articles from abroad in an era without an international copyright agreement allowed American publishers and editors to fill content without paying copyright. Likewise, original writing by U.S. authors was consistently reprinted both nationally and internationally, as in the case of Poe's stories of mesmerism, without remuneration for the author. See McGill 2003.

4. Reese's targeting of "factory girls" also signals how the controversies over mesmerism in both Britain and America were highly gendered. There has been extensive and excellent scholarship on the subject of mesmerism and gender. Of the most recent work, see Daly 2009; Lehman 2009; Mills 2006; Winter 1998; Brown 1990.

WORKS CITED

Albanese, Catherine L. 2007. *A Republic of Mind and Spirit: A Cultural History of American Metaphysical Religion*. New Haven: Yale University Press.
Baily, Pearce. 1951. "The Past, Present and Future of Neurology in the United States." *Neurology* 1:1–9.
Baker, Samuel L. 1984. "Physician Licensure Laws in the United States, 1865–1915." *Journal of the History of Medicine and Allied Sciences* 39:173–97.
Beard, George M. 1877. "Physiology of Mind-Reading." *Popular Science Monthly* 10:459–73.
Beard, George M. 1879. "The Psychology of Spiritism." *North American Review* 129:65–80.
Beard, George M. [1881.] 1972. *American Nervousness: Its Causes and Consequences.* New York: Arno Press and the New York Times.
Brown, Edward M. 1983. "Neurology and Spiritualism in the 1870s." *Bulletin of the History of Medicine* 57:563–77.
Brown, Gillian. 1990. *Domestic Individualism: Imagining Self in Nineteenth-Century America*. Berkeley: University of California Press.
Buchanan, Joseph R. 1842. *Sketches of Buchanan's Discoveries in Neurology*. Louisville, KY: J. Eliot & Co.'s Power Press.
Burman, F. S. 1844. "More Mesmeric Impostors." *Medical Examiner and Record of Medical Science* 7:309.
Child, Lydia Maria. 1998. *Letters from New-York*. Ed. Bruce Mills. Athens: University of Georgia Press.
Clarke, Adele E., Janet K. Shim, Laura Mamo, Jennifer Ruth Fosket, and Jennifer R. Fishman. 2003. "Biomedicalization: Technoscientific Transformations of Health, Illness, and U.S. Biomedicine." *American Sociological Review* 68:161–94.
[Collyer, Robert.] 1845. "Editorial Miscellany." *Broadway Journal* 2:390–92.
Connolly, William E. 2002. *Neuropolitics: Thinking, Culture, Speed*. Minneapolis: University of Minnesota Press.
Conrad, Peter. 2007. *The Medicalization of Society: On the Transformation of Human Conditions into Treatable Disorders*. Baltimore: Johns Hopkins University Press.
Conrad, Peter, and Joseph W. Schneider. 2009. "Professionalization, Monopoly, and the Structure of Medical Practice." *The Sociology of Health and Illness: Critical Perspectives*, 8th ed., 194–200. New York: Worth Publishers.
Crabtree, Adam. 1993. *From Mesmer to Freud: Magnetic Sleep and the Roots of Psychological Healing*. New Haven: Yale University Press.
Daly, Nicholas. 2009. *Sensation and Modernity in the 1860s*. Cambridge: Cambridge University Press.
Damasio, Antonio. 1999. *The Feeling of What Happens: Body and Emotion in the Making of Consciousness*. New York: Harcourt Brace.
Darnton, Robert. 1968. *Mesmerism and the End of the Enlightenment in France*. Cambridge: Harvard University Press.
DeJong, Russell N. 1982. *A History of American Neurology*. New York: Raven Press.
Dods, John Bovee. 1850. *The Philosophy of Electrical Psychology*. 4th ed. New York: Fowler and Wells.

Duffy, John. 1993. *From Humours to Medical Science: A History of American Medicine.* 2nd ed. Urbana: University of Illinois Press.

Freidson, Eliot. 1970. *Professional Dominance: The Social Structure of Medical Care.* New York: Atherton Press.

Fuller, Robert C. 1982. *Mesmerism and the American Cure of Souls.* Philadelphia: University of Pennsylvania Press.

Gauld, Alan. 1992. *A History of Hypnotism.* Cambridge: Cambridge University Press.

Gilmore, Paul. 2009. *Aesthetic Materialism: Electricity and American Romanticism.* Stanford: Stanford University Press.

Hawthorne, Nathaniel. 2002. *The Selected Letters of Nathaniel Hawthorne.* Ed. Joel Myerson. Columbus: Ohio State University Press.

Kuhn, Thomas S. 1996. *The Structure of Scientific Revolutions.* 3rd ed. Chicago: University of Chicago Press.

Lehman, Amy. 2009. *Victorian Women and the Theatre of Trance: Mediums, Spiritualists and Mesmerists in Performance.* Jefferson, NC: McFarland.

Martineau, Harriet. 1844. "Miss Martineau on Mesmerism." *Phalanx* 1:321–23.

Martineau, Harriet. 1845. *Life in the Sick-Room.* Boston: William Crosby.

Massumi, Brian. 2002. *Parables for the Virtual: Movement, Affect, Sensation.* Durham: Duke University Press.

McGill, Meredith. 2003. *American Literature and the Culture of Reprinting, 1834–1853.* Philadelphia: University of Pennsylvania Press.

Mills, Bruce. 2006. *Poe, Fuller, and the Mesmeric Arts: Transition States in the American Renaissance.* Columbia: Missouri University Press.

Otis, Laura. 2001. *Networking: Communicating with Bodies and Machines in the Nineteenth Century.* Ann Arbor: University of Michigan Press.

Pencil, Peter. 1844. "The Mysterious Neighbor." *The Columbian Lady's and Gentleman's Magazine* 1:200–207.

Poe, Edgar Allan. 1984a. "The Facts in the Case of M. Valdemar." In *Poetry, Tales, and Selected Essays*, ed. Patrick F. Quinn and G. R. Thompson, 833–42. New York: Library of America.

Poe, Edgar Allan. 1984b. "Mesmeric Revelations." In *Poetry, Tales, and Selected Essays*, ed. Patrick F. Quinn and G. R. Thompson, 717–27. New York: Library of America.

"Proceedings of Societies." 1841. *Medical Examiner* 4:790–95.

Reese, David Meredith. 1838. *Humbugs of New York.* New York: John S. Taylor.

Rosenberg, Charles E. 1992. *Explaining Epidemics and Other Studies in the History of Medicine.* Cambridge: Cambridge University Press.

Rosenberg, Charles E. 1997. *No Other Gods: On Science and Social Thought.* Rev. and exp. ed. Baltimore: Johns Hopkins University Press.

Sedgwick, Eve Kosofsky. 2003. *Touching Feeling: Affect, Pedagogy, Performativity.* Durham: Duke University Press.

"Seeing Internal Diseases." 1842. *Boston Medical and Surgical Journal* 27: 110–11.

Shortt, S. E. D. 1984. "Physicians and Psychics: The Anglo-American Medical Response to Spiritualism, 1870–1890." *Journal of the History of Medicine and Allied Sciences* 39:339–55.

Silverman, Kenneth. 1991. *Edgar A. Poe: Mournful and Never-Ending Remembrance.* New York: HarperCollins.

Starr, Paul. 1982. *The Social Transformation of American Medicine*. New York: Basic Books.

Taves, Ann. 1999. *Fits, Trances, and Visions: Experiencing Religion and Explaining Experience from Wesley to James*. Princeton: Princeton University Press.

"Triumphs of Mesmerism." 1844. *Medical Examiner and Record of Medical Science* 7:306.

Wilson, Elizabeth A. 2004. *Psychosomatic: Feminism and the Neurological Body*. Durham: Duke University Press.

Winter, Alison. 1998. *Mesmerized: Powers of Mind in Victorian Britain*. Chicago: University of Chicago Press.

Chapter 2

The Performativity of a
Historical Brain Event

Revisiting 1517 Strassburg

Jameson Kismet Bell

Since the early nineteenth century, the military surgeon Hans von Gersdorff's (ca. 1455–1529) *Feldtbuch der Wundartzney* (1517) and medical doctor Lorenz Fries's (ca. 1490–1530) *Spiegel der Artzney* (1518) have been used as a keying mechanism to help delimit the boundaries of the modern brain.[1] Gersdorff's book includes fugitive sheets (*fliegende Blätter*), one of which represents an anatomized body and brain in a single sheet broadside; copies of this fugitive sheet were subsequently reprinted the next year in Fries's text. In the history of the brain, these images mark the beginning of all subsequent, similarly represented cerebral cortices.[2] Hans von Gersdorff and Lorenz Fries were also two of the first to use the combined languages of dissection and typography in defining the style by which doctors and surgeons approached the brain. Contemporary historians of the brain may notice that limited translations of textual fragments from the vernacular German to English, as well as the presence of some of the first visually accurate images of the head and brain, have led both German- and English-speaking scholars to emphasize the images of a dissected head as metonyms for the entirety of Fries's and Gersdorff's knowledge of the brain.

As important as dissection and typography are to a post-sixteenth-century understanding and methods of representing the brain, many histories omit that Gersdorff and Fries were also part of a group of individuals in the early sixteenth century who attempted the difficult work of translating codes of one language into the codes of another, namely, mapping the invisible teleological purpose of the brain onto the visible surface of the printed page. This telos was defined by the cell theory of cerebral physiology in which the instruments of the soul such as common sense, imagination, reason, and memory were believed to be located in the cerebral ventricles. From the fourth through the sixteenth century, the brain was primarily defined by this localization of the mental faculties in the cerebral ventricles, a practice of localization that arguably has lasted in trace form until the present day (Hagner 2006; Kemp 1996).

After 1517, fragments from the languages of dissection and typography—the physical gestures of cutting the body recorded in words created with moveable type and three-dimensional images on the surface of a printed page—established the style by which individuals framed future "brain events." For the purpose of this essay, I define a *brain event* as the conceptual frame that unites brain-related magnitudes of performance. Theater scholar Richard Schechner, author of the well-known *Performance Theory* (1988), argues for a broader classification of an "event" than one that is only limited by space and time. An event, for Schechner, is that which has magnitudes of importance within a culturally accepted unit of time and space. The magnitudes of an event, an idea appropriated from Aristotle's *Poetics*, are the actions that occur as spatial and temporal units combined with these actions' cultural importance, factors that cannot be separated (1954). The magnitudes of a brain event are created by the performative style of combining elements from a multitude of structures within a cultural frame.

As the cerebral cortex became increasingly visible in the sixteenth century, a brain event depended more and more on the measurable volume of cerebral material, eventually contributing to the language for the rituals of mapping the interior and exterior surfaces of the skull and brain found in the practices of cranioscopy and phrenology in the eighteenth and early nineteenth centuries (Hagner 2004), and even in current brain-mapping techniques (Beaulieu 2003). Implied in the process of limiting a brain event to brain matter is the exclusion of those languages that distract one from the measurable and the visible. Measuring and recording tools, the specific individuals using these tools, the setting in which the individuals and technologies measure, as well as the choice of the object(s) to be measured, these often fallible scripts are omitted so that the brain material, and thus a measurable, spatial and temporal brain event, stand out in the historical record. Ludwig Fleck, the early twentieth-century historian and sociologist of science, argued that the recorded gestures of dissection alone could not have produced a "modern brain" in the Renaissance; nor did the increased use of dissection dispel a "medieval brain," if there was such a thing (1979, 35–36). These gestures were simply added to an already existing thought style. In this collection, Murison has shown that understanding this process of exclusion and exclusion in various epochs, specifically the United States in the 1840s, becomes part of understanding the history of the creation of brain and brain events for new audiences.

In addition to Fleck, many scholars have pointed out that there has been a reductionist tendency in the neurosciences to limit knowledge to "that which occurs in the brain" by focusing on a spatial/temporal script that can be visualized and measured (Beaulieu 2003).[3] This essay will use three concepts from theories of performativity, namely, a thought *style* (Fleck 1979), performative *script* (Schechner 1988), and cultural *frame* (Turner 1988), to create an alternate approach to one of the first modern brain events that occurred in early sixteenth-century Strassburg.[4] By reframing the now iconic brain event from 1517 Strassburg, one that has been cited as the origin of modern methods of knowing the brain, this essay will help to show that the magnitudes of a brain event include much more than "that which occurred in the brain." Indeed, we will see that in 1517 Strassburg, Gersdroff and Fries used more than the languages of dissection and typography to approach the brain: they translated the languages of dissection and typography into preexisting technological, epistemic, and moral scripts, and by *script* I mean the smallest recognizable code of behavior, perception, and technology that can be translated from structure to structure (Schechner 1988).

The most important of these scripts was the theory of faculty psychology that localized an individual's mental faculties of common sense, reason, and memory in the cerebral ventricles. These "instruments of the soul" were the technology of the mind, and along with other scripts from poetry, humoral theory, architecture, and social interaction, defined the very possibility of a brain event.[5] Unfortunately, these scripts have been omitted from our histories, since Hans von Gersdorff and Lorenz Fries are often read within the framed context of the modern brain. Reframing the codes used to understand the brain in the sixteenth century—situating dissection and topography in relation to the inner senses—not only will allow us to experience how another culture framed performative languages to create a brain event but will also reveal the epistemic, technological, and social scripts that define the possibilities of our own contemporary and future brain events.

What Was the Brain Event That Occurred in 1517 Strassburg?

Who were Hans von Gersdorff and Lorenz Fries, and what was the brain event that occurred in 1517 Strassburg? Also, how did these authors and their images come to fit within the frame of the modern brain? Taken mainly from Gersdorff's descriptions scattered throughout his *Feldtbuch der Wundartzney*, we can assume that in the winter of 1517, the Bologna-trained physician Wendlin Hock from Brackenau supervised a public dissection in Strassburg. This was one of the first documented public dissections outside of Italy. Hock provided an oral narrative of the dissection; whether it was in Latin or German is not known, though the vernacular German seems plausible since the audience was a nonacademic group of practicing barbers, surgeons, doctors, and physicians (Wieger 1885). From this dissection, the military surgeon Hans von Gersdorff produced his *Feldtbuch der Wundartzney*, a surgery manual that is well known as one of the first vernacular texts to deal with amputations and gunshot wounds (Gurlt 1898). One year later, the medical doctor Lorenz Fries published his *Spiegel der Artzney* as a general medicine text, also the first of its kind in the vernacular German. Both men claimed they wrote their books of surgery and general medicine in the vernacular for the good of the "common man," though the autodidactic implications of this term lacked a larger readership in the sixteenth century (Ölschlegel 1985; Giesecke 1998).⁶ Gersdorff's text was based on the fourteenth-century French surgeon Guy de Chauliac's *Great Surgery* (ca. 1368), and Fries's text was based on the thirteenth-century doctor Arnauld de Villanova's *Mirror of Medicine* (ca. 1308).

Feldtbuch der Wundartzney often contained two fugitive sheets that were copied and printed directly into *Spiegel der Artzney*. One of these sheets featured a skeleton in the form of a popular Death figure (fig. 1). The second represented a cadaver in three dimensions with open abdominal and thoracic cavities (fig. 2). Surrounding this dissected body are a series of seven smaller images providing some of the first representations of the cerebral cortex (numbers 1–6) and the tongue (number 7) to enter printed circulation. The fugitive sheets were created by the artist Hans Wächtlin, a student who learned the technique of representing three-dimensional objects in two-dimensional space in a Basel workshop, a skill influenced by both Albrecht Dürer and Hans Baldung Grien (Haack 1896). In the images found in Gersdorff's *Feldtbuch der Wundartzney*, one can see the printer imprint of "JS" standing for Johann Schott, who also printed Gregor Reisch's *Margarita Philosophica* (1502), an encyclopedic text in which one can find another viscera figure who would be the basis for the now famous

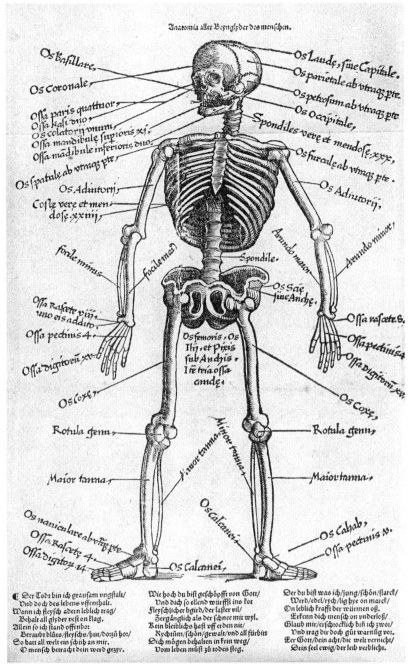

Fig. 1. Fugitive Sheet Skeleton from Hans von Gersdorff, *Feldtbuch der Wundartzney* (Strassburg, 1517). (Courtesy of the National Library of Medicine, Bethesda.)

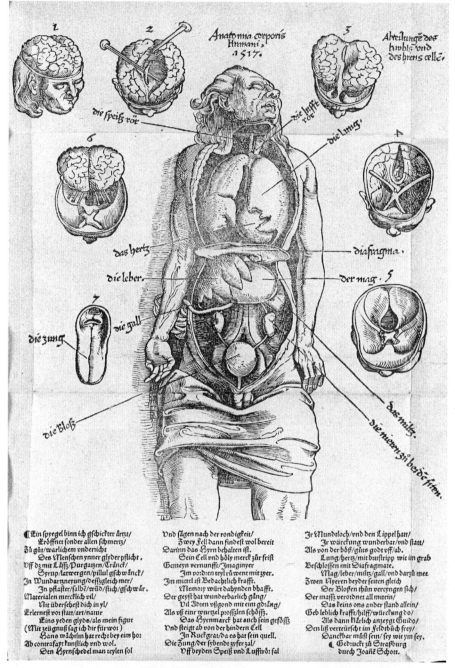

Fig. 2. Fugitive Sheet Dissection Image from Hans von Gersdorff, *Feldtbuch der Wundartzney* (Strassburg, 1517). (Courtesy of the National Library of Medicine, Bethesda.)

fugitive sheet as well as a detailed outline of the cell theory. A rhyming poem (48 lines)[7] often adorns the bottom of the Gersdorff fugitive sheet image, encouraging the readers to trust the veracity of the image as well as directing the reader how to find the invisible instruments of the soul in the ventricles of the brain. The poem begins:

> I am a mirror skilled doctor
> Opened without any pain
> Use for truth to proctor
> Of man's inner parts' domain.

Fries's image of the viscera man, however, was printed directly in his *Spiegel der Artzney* without the clarifying poem (fig. 3). The printer Johann Grüninger appears to have borrowed the woodcut from Schott because of its popularity, commissioning alternate woodcuts for later versions (Sudhoff 1904, 609–10). Both *Feldtbuch der Wundartzney* and *Spiegel der Artzney* were reprinted numerous times throughout the sixteenth century with sometimes slight, sometimes drastic changes to the woodcut images and brain representations.

The history of these images as part of the modern brain begins when Johann Friedrich Blumenbach (1786) mentions them in his *Introduction to the History of Medical Letters* (published in Latin). Ludwig Choulant's *History and Bibliography of Anatomic Illustrations* (1852, published in German), a major contribution to the history of the relationship between images and anatomy, includes Gersdorff and Fries in the transitional section of images created between the Middle Ages and the mid-nineteenth century. Mortimer Frank (1920) later translated, reissued, and expanded Choulant's book from 203 to 435 pages. Choulant's *History and Bibliography of Anatomic Illustrations* argues that through the use of clear language and visually accurate images, one can correctly delineate the true anatomy of the body. He had no idea that the "brain object" recorded in an accurate image and written text would change entirely with the discovery of the electrical excitability of the brain in the 1860s and the nerve cell at the turn of the twentieth century.

The woodcut images themselves have fascinated historians of art and medicine for their accuracy, though the image in Fries's *Spiegel der Artzney* is favored over Gersdorff's, creating an artificial frame for this and subsequent brain images. Writing about the history of anatomical images, Choulant begins his description of the brain by listing only Fries's (Laurentius Phryesen) *Spiegel der Artzney* and the images contained within while relegating Gersdorff's *Feldbuch der Wundartzney* and images to the

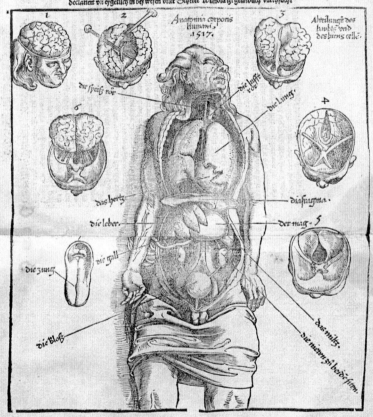

Fig. 3. Dissection Image from Lorenz Fries's *Spiegel der Artzney*
(Strassburg, 1518). (Courtesy of the Bayerische Staatsbibliothek,
Rar. 2053.)

section on fugitive sheets (25–27, 39–42). Frank corrected the error by stating that Gersdorff's book contained the fugitive sheets prior to Lorenz Fries (1920, 128–35). However, rather than include the entire rhyming *knittelvers* (doggerel) poem, Frank offers a detailed description of what a twentieth-century doctor-historian-anatomist would see in the images of the brain that Hans von Gersdorff and Lorenz Fries did not. Subsequent citations of these images also tend to omit the poem from Gersdorff's text in favor of Fries's version that presents the cadaver and brain images sans poem, as well as omitting the tongue (Clarke and Dewhurst 1996, 60). In addition, and because the brain images were originally printed fugitive sheets that could stand alone, any reference to information contained in Gersdorff's or Fries's surgical and medical books is minimal to nonexistent. The neurologist and medical historian Fielding Garrison wrote glowingly that one must recognize the leap forward in observation found in Fries's image (1969), though no mention of the written descriptions of the brain was made.

The omission of Gersdorff's image with the poem in favor of Fries's representation that stands alone further supports the artificial frame between pre- and post-nineteenth-century brain knowledge in which visualization and the ability to measure spatial quantities is essential to what it means to know a modern brain and a brain event. By foregrounding certain script fragments (i.e., the brain images), omitting others (the poem and Fries's and Gersdorff's supplemental information about the brain), and generally being unaware of other cultural scripts (architecture, symbolic objects, and metonymic images, which I will discuss later in this chapter), this brain event emerges as "modern." Bruno Latour has argued that there was never a medieval or modern divide; there are only artificial premodern, modern, and postmodern conceptual borders that need to be ruthlessly defended from real or apparent assailants (1993, 132–35). This defense has assumed the form of performative citationality: scripts that fit the accepted style are repeated through citation, while those that do not are ignored (Butler 1993). Gallagher and Laqueur, following Foucault, have shown the value of historicizing experiences, definitions, and methods of understanding the body: the nineteenth century became the artificial divide between the premodern and modern body (1987). By using theories of performativity to reintroduce myriad technological, religious, artistic, linguistic, and social scripts that went into understanding the brain in early sixteenth-century Strassburg, we will see in the next section that accessing the interaction of these diverse language codes requires an awareness of not only the codes of another culture but also one's own thought style and the limits it places on the possibility of a brain event.

A How-to Guide: Framing, Styling, and Scripting a Brain Event

The following definition will risk repetition and oversimplification for the sake of clarity: a cultural *frame* is the liminal border between thought styles (Turner 1988, 54). A thought *style* is the collection of precepts shared by a group. These precepts determine the future direction and history of a thought collective (Fleck 1979, 64). Repetition of this style, or citation of certain cultural scripts, creates the apparent naturalness of the culturally specific object or event and in turn shores up the frame used by a particular thought collective (30–34). That which is accepted or rejected, the fundamental signs (codes) and their medium of storage and communication, is the cultural *script* (Schechner 1988, 67–73). These three concepts influence and support each other as they help define the performativity of cultural entities such as events, objects, and ideas. Within this frame of performative creation, a brain object will be shaped and contoured as part of a continuum of magnitudes that makes the very brain event possible. The beginning and end of the event, the location, and the fact or fiction of its existence depend on the *frame, style,* and *script.* With regard to Hans von Gersdorff and Lorenz Fries, the addition of script fragments from dissection and typography to the already existing frame of faculty psychology that had defined the brain for more than a thousand years helped create a particular style for understanding brains and brain events in the early sixteenth century.

If we use the theory of cultural scripts to focus our attention, we will begin to understand how a frame and style coordinate these script fragments into meaningful arrangements. For Schechner, a cultural *script* is the basic code of an event, or all that can be transmitted between place and time (1988, 68–71). The *script* is what makes a performance meaningful; it transforms a random happening into something called an "event." As a part of this collection, Fitzpatrick asks the question that has yet to be answered for twenty-first century brain-imaging techniques, namely, "what does brain imaging image?" One should not look to the image itself in order to answer the question. The brain image images the smallest fragments of meaning or the *script* that can be translated to the image for a certain thought style (see Fitzpatrick). For individuals performing the first dissection in early sixteenth-century Strassburg, the *script* was not the printed text that recorded the facts of the event so they could be repeated later. For Gersdorff, Fries, and individuals performing the ritual of cutting a corpse, the script was recorded in the very performance of the public dissection. This performance included many other means of creating, storing, and transmitting codes including very precise gestures from all involved (including the corpse), as well as star alignment (time and date of the dissection), the presence and absence of specific objects and individuals, and architectural designs. The style that combined these code fragments produced a very specific cultural frame in which a brain event was possible.

As a performance theorist, Richard Schechner pays particular attention to how various magnitudes of an event are intimately intertwined. A theatrical performance—such as the public dissection that occurred in 1517 Strassburg—does not commence and end with the actors entering and leaving the arena of performance. Doctors, surgeons, and barbers move seamlessly between epistemological zones, a continuum of performance spaces and times, each with differing magnitudes—street, library, execution of a condemned criminal, procurement of the body, staging a wooden platform for the dissection, the actual theatrical performance of cutting the body before a live audience, postproduction gestures and speech of religious rites for the deceased, burial, and finally recording the important script fragments on an alternative storage medium such as paper (290–322). It would be a mistake to look for the codes of brain event only in printed form, assuming that all the codes of the brain event are stored, retrieved, and combined in printed words and images.

The thought style particular to Fries and Gersdorff recorded brain events in multiple and diverse media such as the Strassburg Cathedral, the space within and surrounding Saint Anthony's hospital for amputations and dissections, the grammar of gestures within the social hierarchy at the dissection, vernacular German and Latin speech, and rhythm and rhyme of a printed poem. The meaning of these script fragments only becomes accessible if approached within the frame by which the movement between the magnitudes is defined in the early sixteenth century, rather than those isolated by measurement and visualization in subsequent centuries.

Schechner's concept of a performative *script* demonstrates that there are different ways to perform and record the codes of an event, particularly the brain event in sixteenth-century Strassburg. Gersdorff and Fries participated in a particular thought style to order these codes into a particular frame, as incongruous as they may seem to us. If one does not recognize the frame and the style in which the code fragments are arranged, the event becomes unrecognizable. Oral codes, gestural codes, written codes, and typographical codes or inscriptions create very different brain events, as is the case with electronic and computational codes (Latour 1988). The particular harmony of script fragments from each of these languages creates and reinforces a particular thought style (Fleck 1979, 38).

So what precisely was the brain in 1517, if not an object in which brain events occur? For Fries and Gersdorff, as well as other individuals in the early sixteenth century, objects such as the brain were endowed with a unique, nonvisible telos or purpose. The cell theory that localized mental faculties of common sense, imagination, reason, and memory in the cerebral ventricles epitomized this telos as well as the expertise needed to access its truth. This thought style, in which an object's metonymic quality symbolized its truth, functioned very well for transmitting codes through oral and gestural communication, yet it proved difficult to represent in print. Prior to 1517, meaningful codes about the brain had been shared and stored through repetition of sounds, gesticulation, and allegorical and symbolic images. As the script to understand the brain changed from oral and gestural to typographical, a medium that restricted the script fragments to infinitely repeatable visual codes, this nonvisible truth could only be represented with much difficulty through mechanically pressed words and images (Bell 2009).

Hans von Gersdorff and Lorenz Fries moved between the spoken word *brain*, the image of the cerebral cortex, and the written word *brain* with much more difficulty than readers steeped in brain images represented in typographic and computational media might imagine. Yet, Gersdorff and Fries moved much more smoothly between analogies, universal truths, spoken words that connect ideas through rhythm and rhyme, gestures, and images that were more important for their symbolic storage capacity than visual accuracy and infinite repeatability. Post-nineteenth-century readers find it difficult to translate the bits of the script from a sixteenth-century brain event into a modern language because they are arranged specific to our thought style. This difficulty also explains the limits placed on sixteenth-century brain events by historians: the brain images in Gersdorff's and Fries's books become the metonymic definition of the sixteenth-century brain because we find it difficult to recognize signs not common to our own language, which prioritizes visualization and measurement. In this collection, Hendrix and May's ability to see a neurological performance in Bernard of Clairvaux (1091–1153) and Teresa of Ávila (1515–82)—like recent sightings of the cerebral cortex on the ceiling of the Sistine Chapel popularized by Meshberger (1990) and continued through Paluzzi (2007) among others—evokes Latour's critique of possible sightings of tuberculosis in ancient Egypt and machine guns in prehistoric times (Latour 2000). Such histories become a means to perform our own knowledge practices so that we create a historical frame by limiting the codes to those that we can recognize or even create; this historical style creates a brain event in the early sixteenth century as an origin story of our own thought style regarding the brain (see Hendrix and May, this volume). As we shall see in detail in the next section, printed words and accurate images are only two of the many magnitudes that Gersdorff and Fries used to define a brain event.

The First Brain Event in Strassburg: Demonstrating a Path to Pith

For Hans von Gersdorff and Lorenz Fries, brain matter experienced during a dissection did not define the location of brain events. The difficulty with the traditional narrative of the "brain event" retold earlier in this essay is that, if one looks alternately at the languages used in the fugitive sheet images, their relationship to the books in which they were attached, and the thought style in which this dissection was performed and understood, Gersdorff or Fries only mention brain matter as that which one must discard in order to understand the idea of the brain. This section will briefly outline the theory of faculty psychology that framed Gersdorff's and Fries's knowledge of the brain as it was applied to defining a brain event. This frame strongly influenced how script fragments from dissection and typography were incorporated into their thought style.

Both Hans von Gersdorff's *Feldtbuch der Wundartzney* and Lorenz Fries's *Spiegel der Artzney* apply the frame of the "inner senses" or "instruments of the soul" located in the cerebral ventricles to dissection and attempt to demonstrate these faculties. Since the fourth century, various epistemologies of the inner senses framed that which could be considered knowledge from that which was sensible, or available to the five external senses. Sensible information had to be translated by the inner senses in order to become knowledge. According to Hans von Gersdorff's and Lorenz Fries's version of the theory of inner senses, the powers or faculties of common sense and imagination were located in the first (anterior) ventricle, judgment and reason in the second (central), and memory in the third (posterior, back of the head). A small worm-shaped valve (vermis) connected the cells and could be controlled though conscious effort. Gersdorff's poem below the fugitive sheet provides an outline of how one was to use dissection to demonstrate this truth by equating an image with a personal witness and visible perception with invisible knowledge (see the appendix for a full translation of the poem).

A personal witness artistic and true
[This image] Tells how one divides the skull in two.
And saw through the roundest section/
Through two skins and you're almost on it
Therein find the brain deposit.
Its cell and holy marrow station
Common sense/imagination
In the front part does lower.
In the center the thoughtful power
Memory would be housed in the rear.

Rational pneuma (spirits) moved between ventricles, with each faculty providing a more refined version of the essence of the object to the respective powers as the spirits moved from the anterior to the posterior cerebral ventricles. The objects of sense differed from common sense; so, too, objects of common sense differed from imagination, which differed from memory. If one imagined movement from front to rear of the head, each faculty provided a more rarefied version of the true essence in question. Importantly, what is largely considered to be metaphor today was at one time believed to be a physiological process.[8]

Early uses of dissection did not attempt to discover new anatomical structures in the brain but to demonstrate the truth of the cell theory, a theory that emphasized a nonsensible *telos* of each object. Often, as exemplified in the theory of the *rete mirabile* mentioned in the poem by Gersdorff (the tiny "miracle net" at the center of the brain that converted blood into animal spirits) dissection demonstrated the existence of nonexistent anatomical structures (Clarke and Dewhurst 1996, 55–59). The brain's telos was to help individuals move from sensible to divine truth, the very definition of the cell theory. This belief in the inner senses that were housed at the core of the brain affected the gestures of dissection: Gersdorff describes a brain that has three chambers or little cells. Each cell has two parts, and each part exercises a power or faculty: "in the first part of the front-most chamber or cell is common sense. In the second part is imagination. In the middle cell is judgment and reason. In the rear most cell is reminiscence and memory" (5r–7v). Even after the belief in the faculty theory had waned, the particular pattern of gestures used to dissect the brain emphasized access and representation of the cerebral ventricles. The script recorded in the gestures remained until Thomas Willis altered the approach to cerebral dissection in the second half of the seventeenth century (Martenson 2004).

Through a study of frontispieces, Andrea Carlino has shown how dissection practices varied between the thirteenth and sixteenth centuries. Thirteenth-century dissection was a performance of a text on the body; this meant that as a teacher read from the ancient authority of Galen, Plato, or Aristotle, a student or barber cut the body to demonstrate the truth of the text, and audience members recognized the truth of the body as outlined by the audible voice of the author (1999a). Early dissections were not intended to provide a tangible investigation of the body, and the poem states clearly that Gersdorff was simply showing the body written by Guido, the medieval French doctor and surgeon Guy de Chauliac (1300–1368). From the gestures outlined in the text and the positioning of participants based on a social hierarchy of doctors, surgeons, students, and barbers, as well as recognizing the organs as their names were read aloud, performing the ritual of dissection was the purpose. The public nature of the dissection ritual implied that individuals were not as much interested in knowing the body as in being seen as they participated in the demonstration of the divine truth of the body (Carlino 1999b).

The elaborate rituals of dissection were intended to mediate between the residue of the material body and the truthful knowledge to be gained. This separation of rational versus sensible knowledge could be found in the very definition of the inner senses and the rituals used to understand, treat, and even enhance brain events the inner senses produced. For example, both Fries and Gersdorff followed the accepted social rules about touching a corpse. Fries, as a medical doctor, possessed a higher social rank than a barber or surgeon. As a theoretically trained doctor, he would only rarely perform a bloodletting procedure and would not participate in a public dissection, even if the purpose were to discover pathology (Siraisi 1990). Lesser-trained barber surgeons performed bloodletting; the fugitive sheets attached to Gersdorff's and Fries's books even emphasize the veins and arteries on the right arm of the corpse, presumably to assist surgeons and barbers in their work, and a "Lassman" or phlebotomy figure is provided in *Feldtbuch der Wundartzney*. Section two of Gersdorff's book demonstrates how and with what instruments a surgeon should treat a cranial and cerebral trauma: images of fractured skulls and trepanation instruments, along with poems describing their use, were didactic tools for a military surgeon, not a doctor. Even though he was in favor of using dissection to demonstrate body knowledge, Gersdorff also states that he believes his duty was part punishment for the criminal, as cutting the corpse was still punishing a "criminal body" in order to redeem his soul.

From this information about social interaction in Strassburg, the images of the dissected cadaver and brain attached to Fries's book seem to be inserted in the text by the printer Johann Grünninger rather than Fries.

It appears that the modern "brain event" was the result of the ingenuity of a printer rather than an advance in medical knowledge.[9] The popularity of Gersdorff's text printed in 1517 must have caught the printer's eye, and he borrowed the woodblock to stimulate sales for Fries's book, published a year later. The only reference Fries makes to dissection and anatomical knowledge is that "some ancients did it" and "dissection only occurs in large cities and is not useful for a doctor," and "see the other man's book in Strassburg who wrote about the third instrument of medicine [a reference to Gersdorff and surgery]."[10] For Gersdorff, the audience, the scalpel, the dissector, the architectural space, the rehearsed gestures, spoken words from differing languages, colors, sounds, smells, time of year, and the theory of the inner senses were all part of the cultural script that defined a brain and a brain event in sixteenth-century Strassburg.

Translating the diverse language scripts of living individuals into typography proved very difficult since equivalent typographical codes capture only part of the oral, gestural, and aural communication. Fries, however, did have a language to define brain events. Using a combined theory of humors, elements, and mental faculty function, Fries could know and influence what happened in the brain. For Fries, brain events in sixteenth-century Strassburg were based on organizing cultural scripts—objects, colors, sizes, temperatures, times, individual and group gestures, and, remarkably, locations—into a particular order based on his thought style. Too much or too little of one of the four humors (blood, phlegm, yellow bile, black bile) as well as one's body heat caused a localized disruption of the animal spirits, which then disrupted the specific mental faculty. Too much or too little of one humor in the front of the head caused difficulty with common sense and imagination; too much or too little of another humor in the center of the head affected reason; too much or too little of another in the rear affected memory.

Fries attempted to harmonize the body's four humors and four complexions with the four elements in nature. Even further, buildings in Strassburg, such as the Strassburg Cathedral, were utilized by Fries to improve cerebral functioning in his patients. Individuals in the sixteenth-century Strassburg Cathedral could be seen rocking forward and backward to stimulate the humors in the head and control the vermis, which would improve the movement of animal spirits through the cerebral ventricles. A short tract on memory authored from 1523 continued the reference to architecture and geography, in which Fries advocated a particular use of one's memory faculty in order to store and retrieve information placed around the city, which later became a reciprocal practice where a metaphor of the brain guided how a building or city ought to be organized. This constructed space in turn dictated how one performed the brain and created brain events (Sennett 1996).

Conclusion

As the concept of the brain becomes increasingly important to styles of twenty-first-century thought, the so-called neuroscientific turn of the last twenty years seems to be a reframing of how one asks certain questions of the brain and about the authority of those asking the questions, and how one expects the brain to respond to those questions. In this collection and elsewhere, Whitehouse emphasizes the *storybank* that is part of the construction of the brain, which in many cultural domains means that stories that accumulate over time trump data. Beyond the verbal story are the imagistic and gestural stories repeated through generations and told through a harmonious movement between elements of each structure. We have seen that the boundary of a "brain event" is the conceptual frame that unites brain-related magnitudes of performance. As such, dissection and typography are two magnitudes among many that have defined the ways individuals approach the modern brain and create a modern brain event. Isolating an event in the brain also implies one can isolate it in a parallel but alternative media (i.e., the book). The repetitive citation of these language scripts is particular to post-nineteenth-century historiography and does not originate in the early sixteenth century.

Through reframing the brain event in 1517 Strassburg, and attempting to translate one thought style into another, new questions have arisen that will be helpful in defining the traces of the historical brain in a post-typographical brain event: is the brain event recorded in the scripts of speech and gesture the same as the brain event recorded in moveable type, printed images, and digital code? Hendrix and May show that in relation to fashioning the self and world, "naturally, different technologies yielded different results." Yet, can one go so far as to say differing technologies create entirely different objects? The *neuro* prefix so common in many brain-related activities today signifies a changing performance of knowledge as neurosciences are increasingly used as a means to understand cultural performances—not just in neurology but also in psychology, psychiatry, education, law, and so on. Like the "inner senses" in the sixteenth century, the *neuroscientific turn* is creating the brain within a particular epistemic frame.

The larger magnitude of the "modern brain event" that arbitrarily began in 1517 can be said to have ended in the year 2000 after the "decade of the brain" translated the typological brain into digital code. Similarly, the discovery of the neuron and electricity transformed the brain at the turn of the twentieth century, and dissection and topography at the turn of the sixteenth century. For Fries and Gersdorff in the early sixteenth century, as well as contemporary humanists and neuroscientists, approaching the brain requires the difficult work of harmonizing codes from multiple languages; translating these diverse scripts also requires an openness to the performativity of a brain event that is spatial, temporal, and culturally meaningful.

If we return one last time to Gersdorff's image and poem (fig. 2), the addition of the representation of the tongue—the seventh numbered figure, the source of language, the divide between the head and the heart, and the organ that "helps the good/evil go up and down"—was an attempt to bring the vivacity of an oral code into the motionless, visual medium (see the appendix for a full translation of the poem). This performative gesture points to the necessity of reading the poem aloud in order for the reader and audience to experience the meaning stored in the rhythm and rhyme of the script. In the same manner, recognizing a brain event created by cultural scripts requires familiarity not only with the brain but also with the particular thought style that gives the event meaning.

APPENDIX

Feldbuch der Wundartzney (1517), von Gersdorff
Jameson Kısmet Bell
Translation of poem from anatomical fugitive sheet (1).

Verse 1

Ein spyegel binn ich gschickter ärtzt/
Eröffnet sonder allen schmerz
Zu gut/ warlichem undericht
Des Menschen ynner glyder pflicht.
Uff dz mit Lässz/Purgazen/ Tränck/
Syrup/latwergen/pillul geschwänck/
In Wundarzneyung/desszgleich mer/
In pflaster/ salbe/ wiid/ stich/ gschwär.
Materialen mercklich vil/
Nit übersehest dich in yl/
Erlernest vor statt/ art/ natur
Eins yeden glyds/ als mein figur
(Mit zeügnuß sag ich dir fürwor)
Hans wächtlin hat recht bey eim hor:
Ab contrafayt kunstlich und wol.
Den Hyrnschedel man teylen sol.

I am a mirror skilled doctor/
Opened without any pain
Use for truth to proctor
Of man's inner part's domain.
From these with letting/purging/potion/
Syrups/confection/pills to gulp/
More in wound medicine solutions/
In plasters/salves/herbs/stitches/expel.
Materials clearly many/
Neglect not your health/
Learn of place/art/nature
Each body limb/ like this figure
(I testify to you the truth)
Hans Wächtlin's privilege is the image here:
A personal witness artistic and true
Tells how one divides the skull in two.

Verse 2

Und sägen nach der rondigkeit/	And saw through the roundest section/
Zwey fell dann sindest wol bereit	Through two skins and you're almost on it
Darinn das Hyrn behalten ist.	Therein find the brain deposit.
Sein Cell und höly merck zür frist	Its cell and holy marrow station
Gemeyn vernunfft/Imaginyer	Common sense/imagination
Im vordren teyl rüwent mitzyer.	In the front part does lower.
Im mittel ist Bedachtlich krafft.	In the center the thoughtful power
Memory würt dahynden bhasst.	Memory would be housed in the rear.
Der geyst hat wunderbarlich gäng/	The mind has wonderful ways/
Vil Adren usgond mit eim gdräng/	Many arteries through which to jostle/
Alls uß einr wurzel prosszlen schössz.	From the root of the porcelain womb.*
Das Hyrnmarck hat auch sein gefössz	Therein does the brain stem dwell
Und steit ab von der hindern Cell	And leave from the back most cell
In Ruckgrat/da es hat sein quell.	The spine/there also has its source.
Die Zung/ der sybendt zyfer zal/	The tongue/the seventh numbered figure/
Uff beyden Speiß und Luffttrör fal	Falls on both air and feeding tubes

*rete mirable

Verse 3

Ir Mundtloch/ und den Lippel hat/	Has a mouth hole/ and also lips around/
Ir würckung wunderbar/ und statt/	Its workings wonderful/ and learned/
Als von der böß/ güts godt uff/ab.	How to make good/evil go up/down.
Lung/herz/ mit brustripp wie im grab	Lungs/heart/with breastbone from the tomb
Beschlossen mit Diafragmate.	All secured by the diaphragm.
Mag/leber/milz/gall/ und darzü mee	Stomach/liver/spleen/gall and even more
Zwen Nyeren beyder seiten gleich	Two kidneys same on either side
Der Bloßen thün vereyngen sich/	That the action of the bladder reunites/
Der massz verordnet all mitein/	The measure orders all within/
Das seins ons ander stand allein/	That neither one nor the other alone can win/
Geb leblich krafft/hilff/würckung do/	But give together life/help/and proper effect
Als dann klärlich anzeygt Guido/	As clearly shown by Guido/*
Den liß verteütsch im feldtbüch frey/	It's been Germanized freely in this Field book/
Dackbar müst sein/ sey wie ym sey.	Thanks must be given, be to him who it be.
Gedruckt zu Straßburg	Printed in Strasburg
Durch Joane Schott.	By Johann Schott

*Guy de Chauliac

NOTES

1. The English translation of Hans von Gersdorff's *Feldtbuch der Wundartzney* is "Fieldbook of Surgery" and Lorenz Fries's *Spiegel der Artzney* is "Mirror of Medicine." All translations, unless otherwise noted, are my own.

2. As is often the case, fugitive sheet images have difficulty surviving historical

durations through degradation or disappearance. The images used in this chapter are found in the 1528 edition at the National Library of Medicine.

3. This process of framing the object or event one wants to discover is part of the practice of science. Lorraine Daston and Peter Galison (2007), building on a generation of scholars' work, have shown that scientific and medical practices are value oriented, and these values or virtues have redefined how objects are created, experienced, and understood (34–35). Historiographically, one can also explore the practices, or the enactment of normative codes (scripts) that create objects and their subjects, rather than focus solely on the objects themselves (39–42). Other authors have shown how methods of measurement and visualization became the dominant scientific style from the seventeenth century forward, not just in neuroscience (Foucault 1994; Duden 1991). This changing style of approaching the world also influenced one's view of the history of the brain; historians in the nineteenth century set boundaries on what should and should not be included in a representation of a brain, as well as the brain events to be studied. These boundaries constrict our definition of a brain event.

4. For consistency, I will use the sixteenth-century spelling *Strassburg* rather than the *Strasbourg*.

5. This theory of the inner senses (often called the cell theory or ventricular theory) defined how fragments from the languages of typography and dissection could be incorporated to help define and explain a brain event.

6. *Feldbuch der Wundartzney* is the only publication attributed to Hans von Gersdorff, whereas the polymath Lorenz Fries had a fifteen-year career of publications focusing on diverse topics spanning astrology, cartography, syphilis, and memory.

7. See the appendix to read the poem in its entirety (in German); Kismet Bell also provides an English translation.

8. Since a detailed outline of the history and use of the faculty theory is beyond this essay, I refer the reader to seminal articles by Sudhoff (1913), Bruyn (1982), and Green (2003).

9. The majority of the images in Fries's text can be traced to one of Grünninger's previous printed works and have no reference to the content of Fries's text (nor are they referenced by Fries). Ironically, some of the images in Fries's *Spiegel der Artzney*, unbeknownst to Fries, stem from Hieronymus Brunschweig's (1450–ca. 1512) *Buch der Cirurgia*, or Book of Surgery, also printed by Grünninger.

10. The three instruments of medieval medicine were regiments (diet), pharmacology, and surgery.

WORKS CITED

Aristotle. 1954. "Poetics." In *Collected Works*, translated by J. A. Smith, 600–628. Oxford: Oxford Press.
Beaulieu, Anne. 2000. "The Space inside the Skull: Digital Representations, Brain Mapping and Cognitive Neuroscience in the Decade of the Brain." PhD dissertation. University of Amsterdam: The Netherlands.

Bell, Jameson. 2009. "The Brain as Material and/or Idea?" *Historia Medicinae* 1:1–9. http://www.medicinae.org/09.

Bruyn, G. W. 1982. "The Seat of the Soul." In *Historical Aspects of the Neurosciences,* ed. Bynum Rose, 55–81. New York: Raven.

Butler, Judith. 1993. *Bodies that Matter: On the Discursive Limits of Sex.* New York: Routledge.

Carlino, Andrea. 1999a. *Books on the Body: Anatomical Ritual.* Translated by Tadeschi. Chicago: University of Chicago Press.

Carlino, Andrea. 1999b. *Paper Bodies.* London: Wellcome Trust.

Carruthers, Mary J. 2000. *The Craft of Thought: Meditation, Rhetoric, and the Making of Images, 400–1200.* Cambridge: Cambridge Studies in Medieval Literature.

Choulant, Ludwig. 1852. *Geschichte und Bibliographie der Anatomischen Abbildung.* Leipzig: Rudolph Weigel.

Choulant, Ludwig. 1920. *History and Bibliography of Anatomic Illustrations.* Edited by Mortimer Frank. Chicago: University of Chicago Press.

Chrisman, Miriam Usher. 1982. *Lay Culture, Learned Culture, Books and Social Change in Strasbourg, 1480–1599.* New Haven: Yale.

Clarke, E., and K. Dewhurst. 1996. *An Illustrated History of Brain Function.* 2nd ed. San Francisco: Norman.

Cunningham, Andrew. 1997. *The Anatomical Renaissance: The Resurrection of the Anatomical Projects of the Ancients.* Aldershot: Scolar Press.

Daston, Lorraine, and Peter Galison. 2007. *Objectivity.* Cambridge, MA: Zone.

Duden, Barbara. 1991. *Woman beneath the Skin.* Cambridge: Harvard.

Eisenstein, Elisabeth. 1979. *The Printing Press as an Agent of Change.* Vols. I, II. West Hanover: Cambridge University Press.

Fleck, Ludwig. 1979. *Genesis and Development of a Scienific Fact.* Chicago: University of Chicago Press.

Foucault, Michel. 1970. *The Order of Things.* New York: Random House.

Foucault, Michel. 1994. *Birth of the Clinic: An Archaeology of Medical Perception.* Translated by A. M. Smith. New York: Random House.

Fries, Lorenz. 1518. *Spiegel der Artzney.* Strassburg. http://daten.digitale-sammlungen.de/bsb00025948/image_1. Accessed on 01.24.2012

Fries, Lorenz. 1523. *Ein kurzer bericht, wie man die gedechtniß wunderbarlichen stercken mag, also das ein yed in kurtzer weil geschrifftreich werden mag.* Strassburg. http://daten.digitale-sammlungen.de/bsb00023838/image_1. Accessed on 01.24.2012

Gallagher, Catherine, and Thomas Laqueur. 1987. *The Making of the Modern Body.* Berkeley: University of California Press.

Garrison, Fielding. 1969. *History of Neurology. Rev. and Enl. with a Bibliography of Classical, Original and Standard Works in Neurology.* Edited by Lawrence C. McHenry Jr. Springfield: Thomas.

Giesecke, Michael. 1998. *Der Buchdruck in der Frühen Neuzeit.* Frankfurt am Main: Suhrkamp.

Green, Christopher D. 2003. "Where Did the Ventricular Localization of Mental Faculties Come From?" *Journal of History of the Behavioral Sciences* 39:131–42.

Gurlt, E. J. 1898. *Geschichte der Chirurgie.* 3 vols. Berlin.

Haack, Friedrich. 1896. "Wechtlin, Johann." Vol. 41, in *Allgemeine Deutsche Biog-*

raphie, 369–71. http://www.deutsche-biographie.de/pnd119865181.html. Accessed on 01.24.2012

Hagner, Michael. 2002. *Geniale Gehirne: zur Geschichte der Elitehirnforschung.* Göttingen: Wallstein.

Hagner, Michael. 2006. *Der Geist bei der Arbeit: Historische Untersuchungen zur Hirnforschung.* Göttingen: Wallstein Verlag.

Kemp, Simon. 1990. *Medieval Psychology.* New York: Greenwood Press.

Latour, Bruno. 1987. *Science in Action.* Cambridge: Harvard University Press.

Latour, Bruno. 1988. "Visualisation and Cognition: Drawing Things Together." *Knowledge and Society Studies in the Sociology of Culture Past and Present* 6:1–40.

Latour, Bruno. 1993. *We Have Never Been Modern.* Translated by Catherine Porter. Cambridge: Harvard University Press.

Latour, Bruno. 2000. "On the Partial Existence of Existing and Nonexisting Objects." In *Biographies of Scientific Objects*, 247–69. Chicago: University of Chicago Press.

Martensen, Robert. 2004. *The Brain Takes Shape: An Early History.* Oxford: Oxford University Press.

Meshberger, F. L. 1990. "An Interpretation of Michelangelo's Creation of Adam Based on Neuroanatomy." *JAMA 14*:1837–41.

Ölschlegel, Rudolf Christian Ludwig. 1985. *Studien zu Lorenz fries und seinem "Spiegel der Arznei."* Dissertation. Barth-Ostsee. R.C.L.

Paluzzi, Allesandro, Antonio Belli, Peter Bain, and Laura Viva. 2007. "Brain 'Imaging' in the Renaissance." *Journal of the Royal Society of Medicine* 100, no. 12: 540–43.

Saenger, Paul. 1997. *Space between Words: The Origins of Silent Reading.* Stanford: Stanford University Press.

Sawday, Jonathan. 1995. *The Body Emblazoned: Dissection and the Human Body in Renaissance Culture.* London: Routledge.

Schechner, Richard. 1988. *Performance Theory.* London: Routledge.

Sennett, Richard. 1996. *Flesh and Stone: The Body and the City in Western Civilization.* New York: W. W. Norton.

Shortt, S. E. D. 1984. "Physicians and Psychics: The Anglo-American Medical Response to Spiritualism, 1870–1890." *Journal of the History of Medicine and Allied Sciences* 39:339–55.

Shortt, S. E. D. 2003. "Brains, Maps and the New Territory of Psychology." *Theory and Psychology* 13:561–68.

Siraisi, Nancy. 1990. *Medieval and Early Renaissance Medicine.* Chicago: University of Chicago Press.

Sudhoff, Karl. [1904.] 1961. "Lorenz Fries of Colmar." *Allgemeine deutsche Biographie & Neue deutsche Biographie.* Berlin. http://daten.digitale-sammlungen.de/bsb00008407/image_772. Accessed on 01.24.2012.

Sudhoff, Walther. 1913. "Lehre von den Hirnventrikeln in textlicher und graphischer Tradition des Altertums und Mittelalters." *Archiv für Geschichte der Medizin* VII, no. 3: 149–205.

Turner, Victor. 1988. *The Anthropology of Performance.* New York: PAG.

von Gersdorf, Hans. 1517 (1528). *Feldtbuch der Wundartzney.* Strassburg: Johann Schott (Rare Books Collection, National Library of Medicine).

Wieger, Christoph. 1885. *Geschichte der Medizin und ihrer Lehranstalten in Strassburg.* Strassburg: Trübner Verlag.

Chapter 3

The Neural Metaphor

Kélina Gotman

This is why the meaning of a metaphor cannot be reduced to its "true" referent: it is not enough to point out the *reality* to which a metaphor refers; once the metaphorical substitution is accomplished, this reality itself is forever haunted by the spectral *real* of the metaphorical content. (Slavoj Žižek, *The Parallax View* [2006], 169)

When Descartes complained that the Ancients had misunderstood the nature and quality of the passions of the soul, in his work of the same name (1649), he drew on the brain's mechanisms, as well as the fibrous ligaments called nerves, to describe the motion of the body, soul, and passions. He argued that whereas it appeared that the soul moves the body to act, and injects it with animal spirits, in fact the body moves of its own accord, through various mechanical operations effected by the nerves, the muscles, and the brain; and that the soul, whose thoughts could be deemed actions and whose passions were translated or transposed into the body's machinic system as tears, laughter, or fear, communicated with the body-machine through these passions. These passions connected the body and mind. They were useful in ethical and political matters, moreover, awakening in the person a feeling of virtue, anger, disdain, or pride, stimulated in theatrical events, for example, moving the person to act. The tenuous site within which these connections between body and mind were performed was called the pineal gland, a small organ situated at the nape of the neck, between the brain and spinal tissue (Descartes 1996, 99ff).

This chapter argues that Descartes's conciliatory operation—between the machinic body and the thinking mind, as well as between emotions, politics, and society—can be placed in a long line of philosophical works

mapping neural, physiological, and animal functions onto social, philosophical, and political domains, often through metaphor. It further argues
that the particular quality of the "neural metaphor" offers a seductive
portrait of society and human life: networked, changeable, full of flows
of information and capital and goods, conveniently biological as well as
subject to a form of free will. And full of imitations and counterimitations,
the superbly ancient image of the mirror underscoring our most often
repeated (and overtheorized) type of interpersonal relationship: mimicry.
The contemporary turn toward the neurosciences, in this light, suggests
a disembalancing of the mimetic relationship, from communication and
exchange toward objective truth-value on the one hand (the scientist is
always, in the end, "right"), and myth, image on the other (the artist is an
idealist, a dreamer, still, ineluctably, "disconnected"); and yet, the force of
image in these idealizations of our networked, interconnected, and flowing
selves pits science and art in the same ball game: making sense of our sensations, like Descartes made sense of his in his famous sitting room. Only
now we're looking at each other. Through this, the "neuroturn" offers a
convenient, evocative first principle, an object small enough to be nearly
invisible (almost abstract, infinite), and massive enough to encompass and
fuel the human machine. In this sense, we are still pre-Socratic: if it's not
water, or air, it's "neurons," refracting and reflecting onto one another, jos-
tling for an aisle (or a window) seat, through and in spite of us, as we try to
catch up and crack their code.

The role of the neural metaphor in this view offers a "networked" conception of actions performed intra- and interindividually, at a metaphoric
level as well as at the level of "real" biosocial processes (verifiable through
scientific method, that is, laboratory work and experimentation, sanctified
by institutional approval). What I am describing as the "neural metaphor"
shapes the biological and sociopolitical models it seeks to describe as much
as it is shaped in turn by them, in a rhetorical to-and-fro between science
and culture, scientific modeling and philosophical analysis.

My aim is to think through how productively *literary* our modes of intellection may be in regard to self-conceptions and philosophical models, both
creatively to reconcile apparently opposing views of society and human
"nature"—through the fluid and malleable, as well as intuitive, image of the
neural pathway, the synapse, or the mirror neuron—and vividly to illustrate
human behaviors through the satisfyingly productive synthesis between
scientific and philosophical thought offered in this metaphor.

This synthesis between science and philosophy—an ancient one, following Latour, who reminds us that we have "never been modern," inasmuch

as we have never quite totally believed in scientific rationality or knowledge afforded us by deductive method, laboratory work, and science—nonetheless offers a contemporary view of science and the humanities as being additive, and complementary. We don't need to choose between science and culture, since science offers such a compelling image of culture and society refracted and reflected through the modes of operation we already know: the complexity of traffic jams and computers, hacker networks and terrorist ones. These make up the stuff of contemporary life, and map superbly onto the neurological models offered up to us from the scientific trenches. Not that these wouldn't in turn be informed by scientists' own trajectories through traffic jams and hypertext, newsreels and flash mobs. As Canguilhem has noted, scientific ideas are always fundamentally informed by the world in which scientists work and through which they generate their scientific ideas (2000, 62). That is not to say that these ideas are purely contingent upon the manifest reality of the world "out there"—or dependent upon the appearance of daily life (and historical condition)—but that they tend to reflect and to refract truths (and phenomena) manifest here and now. If the theory successfully describes some aspect of life as it is lived and experienced, even through counterintuitive operations contradicting the appearance of reality, then perhaps they will continue to do so for a while: they will outlive their context and place of germination. This is the scientific ambition: an image of reality transpiring through models persisting over time.

But, ideally, the congruence of these humanistic and scientific modes of apprehending "bare life" conveniently correlate, offering an image of a greater certainty, and a higher or thicker truth, in this compounding. It is as if we could say: because x and y have recognized the same thing (or a similar one), the chances that it is true are greater. This is not a very scientific principle, but it may help to explain the exuberance and enthusiasms for the neural today among disparate intellectual operators across fields. The arts and humanities in particular can be especially comforted in the emergence of an apparently graspable, as well as verifiable intellectual paradigm sanctified by science, still—in the indubitably scientific age—master of all things "really true," even more so now that we are returning back on the poststructuralist turn to hard facts pleasantly lining up with the philosophical relativism of the last fifty years.

The particular tenor of the neural—as synapse, mirror, flow, network, and fundamentally as interaction—perpetuates the postmodern and poststructuralist projects of additive modes of intellection; reconciliation between opposites; syncretism; connectivity; juxtaposition; and heterogeneity (over purism, simplicity, reduction); just as it seeks to regain confidence in human and social processes by linking these back to an evanescent biological core, away from the relativism and uncertainty of the very same postmodern and poststructuralist tropes whose structures it mimics. This post-postmodernist or post-poststructuralist twist on the "two cultures" divide (between science and society, or science and art) gains the best of both worlds: at once a comforting anchor in the "neuronal Real" and the familiar flows and shifting terrains of Deleuzian poststructuralism: a new intra- and interterritoriality reassuringly mapped onto scientifically proven (or provable) fact.

The Operation of the Metaphor: Arendt, Deleuze, DeLanda, and Capital Flows

The operation of the neural metaphor is ambivalent: at once a purveyor of scientific goods, and a delightfully rich complex of images, morphing and pulsating through an extended body politic, it offers the myth of reconciliation. It confirms poststructuralism (Deleuze was right after all: we live in a world of flows), and it grounds the at-times-dizzying quality of this poststructuralism (are we really all a "we"?) into the hard world of science and laboratory. Neuroscientists busily working away at decoding the processes governing smell, taste, memory, imitation, love, desire, fear, anger, cravings, movement, thinking, inventing, creating, math, maps, and visions of ecstasy tell us that we're all in our brains, and our brains are communicating with our limbs, and other people's limbs, and that it all *makes sense*. But at the same time, all these things that they tell us suggest a fundamentally complex view of human and animal life full of pathways, and full of choices. These underground burrows that sound and look uncannily like roots and rhizomes as described by Deleuze and Guattari, and the movement of bees in a social group working at finding the honey as described by DeLanda, resonate with the philosophical landscape of the post-1968 generation that has suffused the world of arts and humanities thought.

Assemblages and complex structures allow us to make a revolution by gathering together in masses (molecular or molar ones, in Deleuzian terms). But they also recognizably look like objects and structures explainable through biosocial processes reducible in their turn to something that may resemble a cell, or a mirror neuron (the ultimate graspable, translatable, transposable image), or a neural code sharing complex information across multiple bodies. But why resort to the metaphor? We might call it a metaphor complex: the neural is more than just what neuroscience tells us about who we are and why we do what we do, think what we think, and believe in the ways that we might. The neural metaphor offers a complex of images and structures that foreground principles of flow and communication, whether between two or between many, or within one, turning the debate on identity into a pre-Socratic agora, a One and a Many conjoined through mutual operations (and pheromones).

The counterargument is an easy one: humanists, in overrelying on the neural metaphor, a seductive, protean image to describe every sort of conjunction, synapses ready to tell us the story of all our desires, our mood shifts, and political processes, depend upon scientific discovery, and on science's "objective" truth. But the metaphor complex of the neural (or any other) is as much subordinated to scientific thinking and imagery as it steals creatively from it—as it would from any other field (geography, mathematics)—for sheer pragmatic purposes: poetically, evocatively to describe the world that we know (the reality of all these currents, rhizomes, and microexchanges). And, significantly, scientific thinking itself has borrowed from enough images across fields for any "two cultures" jealousies to subside. The image of the mirror neuron itself is highly literary, metaphoric even, arguably because we can best apprehend concepts through image, and it is these images that help ideas to stick or to spread.

But the history of the biologic metaphor in the human and social sciences merits another side-glance, for its darker history, and ethical dimensions. Hannah Arendt argues, in *On Violence* (1969), that "nothing, . . . could be theoretically more dangerous than the tradition of organic thought in political matters by which power and violence are interpreted in biological terms" (75). Justification for violence on biological grounds is tantamount, for her, to a physician imposing a surgical operation on a patient who might be treated noninvasively. The "glorifiers of violence," she argues, in appealing to "organic metaphors," are falsely and dangerously positing human violence as a natural human condition (1969, 75). The distinction she presents between political and social processes that include nonviolence and pacific approaches to conflict, and the social Darwinist approach to society that justifies war on biological grounds, crystallizes antiwar anxieties about claims to a violent biological (and organic) core. What this argument suggests is that the rhetorical operation of the metaphor performed as a politically invested form of intellection dangerously describes a society considered, in the scientific view (twisted by political pundits or social agitators), as inarguable. The neurological grounds for violence, similarly, sanctify animal behavior as if it were just that: imprinted into our sinews and bones.

But the metaphor can work both ways: with the neural, it also recuperates and reappropriates the biological to describe society as movable, dynamic, flexible, and deeply social (in the Tardian sense of being *inter-individual*). This is the neural metaphor mobilized to emphasize flows, and, conveniently, the constant flow of interindividual movements including capital in all its forms (cultural, social, and material).

For post-Deleuzian theorist Manuel DeLanda, the "organismic metaphor" (2006, 8) describes the dynamically biological way in which society operates as a large, messy, vital body firing in every direction. Assemblages of individuals operate as so many cells in a giant organism; the individual body echoes the operation of groups of bodies, and vice versa. More subtly, the old medieval conception of the social and urban fabric as a giant body, with its organs—heart, head, liver, and the flows of humors and animal spirits—finds a new lease on life in the neural metaphor mapping synaptic flows onto urban terrains, full of streetlights and electric currents, roadways, computer networks, and city traffic (including the aptly named information highway).

If the metaphors for neural processes and neural coding obtain their structures from the deep biologic systems neuroscientists are laboring to decipher, the neat formal mapping of one network of flows (capital, goods, desire) onto another (cells, nerves, gray matter) offers a more powerful image of reality than any architectural or social theorist needs to disavow. Enough to recognize the wonderful parallels in all these systems: honey bees, herds of cattle, cells in the body, humans in Shanghai or New York. All of these assemblages working busily away acting and interacting on, through and with one another according to relations of multidirectionality, and a microcausal system of exchange, offer micromodulations preserving overall stability, within the perennial Heraclitean flow. Nature and culture are wrapped into one, moving, predictable, dynamic, and "naturally" adept at resisting or adopting capital or other forms of exchange in equal measure. This model conveniently both allows for resistance (the information passes or it does not; the switch goes to 0 or 1), and it allows for perpetual microinstances of change. This is a Tardian paradise, recuperated by Deleuze and Guattari in "Micropolitics and Segmentarity," from *Capitalisme et schizophrénie II: Mille plateaus* (1980), in which they note the pioneer sociologist Gabriel de Tarde's premonitory attention to interindividual change. All society, for Tarde, in *Les lois de l'imitation* (1890), is based on micro instances of imitation, micro moments of adoption and adaptation, micro exchanges between individuals making up a subtly shifting whole (Deleuze and Guattari 1980, 267).

Deleuze and Guattari go further with this image of the network, the flow, and the territorial (or reterritorialized, and deterritorialized) passage between instances and portions or segments of individuals—recast as "dividuals," for the ever-incrementally networked ways in which human, animal, plant, and mineral life transforms, shifts, and changes, from one moment and one piece of stuff (a cell, a breath, desire) to another. This old Diderotian conception, from *Le rêve de d'Alembert* (1769), in which Diderot casts in the mouth of the delusional encyclopedist d'Alembert a half-waking rant on the porosity of mineral, plant, and human substances—and their constant interchangeability—borrows its language and tenor from a range of scientific disciplines, including biology, neurology, human geography, and mathematics. The processes of intersubstantial transformation that Deleuze and Guattari, after Diderot, Bergson, Tarde, and so many others, articulate rest fundamentally on a vision of life as being united in constant change: vitalist, Heraclitean. This is not, as Badiou claims, in his retort, from *L'être et l'événement* (1988), because "situations are nothing more, in their being, than pure indifferent multiplicities" (Badiou 2005, xii). It is not sheer difference, sheer interindividuality, sheer multiplicity, or (especially) sheer relativism, but a complex network and system (even an antisystem) of dynamic passageways and modular shifts in intensity redolent of the very neurologically and geologically determined ways in which objects come to be transformed, in and through one another. The processes are not random. They are not, either, for that matter, predetermined. This ambivalence perfectly matches the Deleuzian (as well as Tardian, and Diderotian) conception of micromodular change with advances in the neurosciences, showing scientifically how pheromones, for instance, communicate information invisibly, suggesting flows of desire; or how imitations occur, by way of mirror neurons, through instances of facial and emotional recognition, gestural codes, and bodily or vocal tones.

While the one world of references does not map perfectly onto the other (or vice versa), this preparedness to receive and to process, and in turn to transform, models of scientific and social behavior (approaches to life, history, memory) reinforces the one model's viability in relation to the other. Whereas Deleuze and Guattari, following from early sociology and Enlightenment philosophy (at least in Diderot's iconoclastic formulation), argue for a slipping of disciplines, and poetic juxtaposition and reconfiguration of math, biology, geography, and so many other worlds onto the tropes and questions of philosophical investigation, neuroscience offers a conceptualization of life and social process that formally mimics, maps onto, and agrees with the images that they offer. This happy conjunction of worlds of thought and language enables an exponential adoption of the one or the other, compounded, as I am arguing, by a post-postmodern return to scientific (and scientistic) explanations for human, social, and political life. The forms were set, as reconfigurations of old tropes; and the science comes to ground and to solidify and even, perhaps, awkwardly, to prove them.

The fascination for scientific methodology that has transpired in the arts and humanities in the last ten years—due not least to the completely disproportionate funding opportunities afforded the one scientific world over the other humanities one—has also served to compound these fields in a relatively nonegalitarian relation: science trumps art, by proving it. Science has the last word. But the forms in which this science is adopted (and translated) into arts and philosophical discourse are prepared for by already syncretistic formulations, whose preinvestments in a range of disciplinary worlds offers a deep flexibility and adoptability across any range of scientific turns. The particular image of the network and pathway, the flow, the rhizome, the assemblages and disassemblages, the terrorist networks (aptly called cells) so easily mappable onto neuroscientific language only facilitates and enhances this task. We "all" feel, in the end, that we are speaking the same language. This is the great post-postmodern myth.

The neural in all of this fits seamlessly into a conception of society at a critical juncture beyond postmodernism, where change is constant, and yet there is—conveniently offered by the neurosciences—some reason to all this rhyme. Neurons are firing constantly, offering up a patterned and yet open relationship between genetic coding (the stability of history, the past, the family, and the wherefroms) and individual free will: the ability to decide between a pink and a yellow iPod, a Conservative or a Liberal government; a Big Mac or a stir-fry. All of these decisions are in turn performed through a constant interface between agents—what Latour describes as the "actor-network" paradigm—making decisions and processing, as well as translating, and mediating between micro instances of information. The metaphor expands upon a conception of society as it actually may be at any one time, full of flows of information, an excess of communication, and constant transformation and exchange, while preserving the scientific prejudice inherited from the Enlightenment and institutionalized at every level of media and other social transformers: we flow, we are networked, we are animal, we are fluid, we can rest on scientific knowledge for our self-understanding, while maintaining a form of micromodulating free will. The tentative quality of this worldview is both reassuring and liberating: a perfect post-postmodernist conception, liberated from the strictures of modernist realism and rationalism while benefiting from all the free play afforded to us by Lyotard and the generation of 1968.

I don't mean to suggest that the neural metaphor solves all our problems, relieves us of angst (what is true? what is real? who am I?), or magically dissolves the nature-culture divide. But, as a variation on the old medieval model of the extended metaphor mapping body structures (heart, liver, lungs, nerves) onto social and cultural, political and ethical ones (family, hearth, cities, streets), it affords a conveniently atemporal schema offering reciprocity between worlds. It is at once the "machinic" body (the body as system), and it accounts for social and individual passions; it conjoins these through the image of the neural pathways exchanging information and going about their business creating order through the chaos of cells, reconciling animal and biological determinism and the political fantasy of free will. It also conjoins, significantly, individuality and a dynamic collectivity: we are all cells in the great body of the city, and world. The neural is a "both and" metaphor, bringing together micro and macro operations at the level of science (the neural body) and social and philosophical investments (the "I") simultaneously. Like the best metaphors, in Ricœur's formulation, from *La métaphore vive* (1975), it enhances the world it is meant to describe, while changing it—ever so slightly—in the description.[1]

Psychoanalysis at War with "Brain Science": Freud, Žižek, and the Real

In Žižek's terms, this is a parallax view, shifting the object as the point of view shifts in the labor of apprehending it. The particular shift operated between the "brain sciences" and the Real (opposing, as he enjoins us to do, the neuronal Real with "another Real") (Žižek 2006, 177) offers a view of the world as composed of two (or more) discursive realms: one scientific one, which has an almost imperialist insistence on dominance in the realm of truth (and the Real), and one other one, perhaps a phenomenological one—certainly a psychoanalytic one, in Žižek's conception, which accurately describes sensations as we know them in the day-to-day. Consciousness—what we do when we are thinking about neurons, for example—operates as the locus of the real Real, in which the incursion of the scientific method on practice and philosophical understanding is undone or set aside. This is, for Žižek, psychoanalysis at war with the brain sciences, seeking to regain its own ground, against the terrorizing presumptions of scientific (and scientizing) monopolies on truth, the real, and the mind. In order to counter scientific discrediting of the validity of psychoanalysis, another Real has to be posited. This is the real of precat-egorical, preschematized, or pretranscendental systematizations of reality and experience: the apple before it has become an Apple—part of the Apple category—or pain before it has become construed as a neural network of synapses describing pain as a disorder of, for example, the brain.

Interestingly, the distinctions Žižek is drawing between a neuronal and an other Real (the Real of psychoanalysis, in this instance) echo Freud's anxieties about failing to follow through with scientific methodology after he had started out in the biological laboratory of the University of Vienna in Trieste in his search for the male neuroanatomy of the eel. Freud moved away from neuroanatomical research there and at Charcot's clinic at the Salpêtrière—where he was confronted with early neurological research on male hysteria—toward an increasingly abstracted formulation of the brain processes, from language dysfunctions including agraphia, aphasia, word-deafness, and alexia to the Ego, Id, and Superego that we know. This process of abstraction, effected over decades of research and interrogation into the functions of mind, dreaming, desire, and lack (taken up by his student Lacan, in particular), moved beyond the frontiers of neurological science, as Freud well knew, without, for that matter, wresting itself from the need for scientific justification entirely. In his introduction to the *Studies in Hysteria*, "Case 5: Fräulein Elisabeth von R." (1985), Freud remarked on this anxiety: "It strikes me myself as strange that the case histories I write [here] . . . lack the stamp of science. I must console myself with the reflection that the nature of the subject is evidently responsible for this, rather than any preference of my own" (quoted in Gamwell and Solms 2006, 14). Although Freud began to formulate a theory of the "neuron doctrine" early on in his research, Wilhelm von Waldeyer-Hartz and Santiago Ramón y Cajal would be credited with this discovery nearly a decade after Freud's drawings pointed toward this little bundle of nerve cells and fiber (Gamwell and Solms 2006, 69ff.).

The irony of course does not end there. With the advent of mirror neurons, Lacan's own exploration of the function of imitation (and narcissism) in the image of the mirror—and the mirror stage among infants—would find its scientific validity in the very bundles of nerve cells and fiber that his predecessor had begun to draw out. The extent to which these metaphors, drawings, images, and schematizations of the brain and human behavior help to explain how it is that we grow to hold beliefs or desires is neither here nor there: dancers and other artists have been moving in droves toward the promise of new social validation in the form of neuroscientific research. An electrode in the brain not only funds some art practice (this is the most cynical, pessimistic view), it also sanctifies aesthetics with the stamp of good science (perhaps an equally cynical view). Dancers and actors know mirrors intimately: all of performance practice is generated and sustained before them. Imitating masters, breaking with tradition, finding a form or a shape through self-observation, these are tropes wellworn in the performing arts, among other disciplines. That such a handy little object as the mirror neuron might explain why audiences laugh at comedy, cry at tragedy (we hope), or subtly tense a muscle when watching a fouetté unites, once again, all of our disparate worlds under the banner of the new atom.

It doesn't, however, particularly explain change. Perhaps this is the social conservatism of the image translated and metaphorized into the humanistic and social realm; but perhaps it has also simply failed to explain away the real beyond the neuronal Real. It has failed to account for Žižek's "enigmatic" Real, the "indivisible remainder," where understanding and self-awareness happen outside the scientific realm (Žižek 2006, 177). Perhaps there is a gap, an aporia, between neurology, sociology, and the arts. Perhaps we knew that all along.

But perhaps the consciousness beyond consciousness has simply not yet been discovered, as fast and hard as we humanists hang onto the thought that it can't possibly ever be; and the mirror neuron—as a function of imitation—is not going to help.

Memory, Forgetting, Brains: Ethical Injunctions, Evanescent Images in Ricœur, Virno, Laborit

When I was young, my neurophysiologist father mused often about the paradox of the brain and total knowledge: the brain can never totally know itself, since there is always some piece of the brain left over doing the knowing. This image has stayed with me, recurring in different forms over the years, as if there were some supplement, residue, or some backstage space of pure knowing, a function or a zone occupied with the activity of doing the knowing. An ungraspable region—perhaps a shifting one—an evanescent, impossible, "I." The I before the "I" (and always beyond it). In Ricœur's terms, the politicoethical problem of the "I" as agent—a knowing and remembering agent—places a heavy weight on the task of neuroscientific investigation. If we fail to remember (e.g., the Shoah), is this because we cannot? Or we do not (we dare not)? What is the neurobiology of forgetting? What is the responsibility of scientific investigators to understand this reality? Can they be exempt? Must they continue to claim to be exempt? And to what ends?

Ricœur's claim, in *La mémoire, l'histoire, l'oubli* (2000) hinges upon the problem of the ostensive amorality of scientific research and, as he notes, the paradox of neuroscience's inability to say anything about the moral or ethical conduct of everyday life (548). It is not a question of failing to see but of failing to see how necessary it may be to see: the opacity of science trumping its ethical and political potentiality; a triumphalist amoralism claiming scientific investigation beyond good and evil. Paulo Virno, in his short essay "Mirror Neurons, Linguistic Negation, Reciprocal Recognition," in *Multitude: Between Innovation and Negation* (2008), notes that the whole problem of the mirror neuron rests on the ethicohistorical problem of sympathy: between one human being and another, between an executioner and his victim (169ff). It is the ethical dimension in both these cases that challenges the neural for failing to contend with the consequences or premises of claims put forward, under the banner of pure science.

But, if we imitate one another, how do we account for individual agency, resistance, evil? If we forget, how do we account for the necessity never to repeat? Is this a task for philosophers and politicians across the "two cultures" divide? Or a task for new "neuro-" syntheses, somewhere betwixt and between pure science and the softer nether-regions of the arts? And what if the arts are buried away in their turn in the rich complexity and the beauty of the neural metaphor, sufficient in its right?

Biophilosopher Henri Laborit offers a slightly different take on this conundrum, when he suggests in *Éloge de la fuite* (1976) that we are biologically always already ethically and politically involved with one another, through the very neural pathways that connect us at the level of the "I." We feel pain at the death of a loved one, he argues, precisely because shared memories—a picnic that was rained upon that day in 1954, when grandma pulled her skirt over her head, say, or the day you told me "I love you"—are stored neurobiologically within the very neural makeup of those persons who experienced these moments together. The piece of me that was there that day is shared by those who were as well, and remember it. When my father or my mother dies, they take away with them the days of my earliest youth—those I can only remember through them, and the stories that they tell. A little piece of me dies when they do: the piece I cannot remember. A little piece of me dies when a good friend dies, as well, because those shared moments I no longer remember go with her or him. We are all connected, in this way, to those we love and share time with, neurologically as well as (perhaps more than) "purely" socially, or psychologically. The ethical injunction embedded in this observation compels us to remember that we are bodily contained within one another, insofar as a neural pathway and a memory are "bodily" conjoined across persons: bodily in the sense of a micro object, an entity, a force, a flow—in unequal measure, depending on the amount of time shared or the intensity and density of the complex of emotions in time (Laborit 1976, 8off.).

Paradoxically, this image too—the pain at the death of a loved one, explained through shared moments stored in a complex of neural networks and synapses—has stayed with me, as a story told to me by my father, one day when I was sitting at the back of the car. He denies ever knowing this, or telling it to me, but I'm sure—I'm certain—that I didn't make it up. I hadn't known Laborit's work before.

Bergson's Dream: The Myth of Pure Presence, the Neural Body as Flux

The mysteries of these moments and acts of forgetting are powerfully deployed in the work of metaphorizing pictures of the mind and body organized in time—through vital mechanisms—defining us as individuals and as collectivities, operating principally through "flux." Bergson reconciles objects in the world and their representation in *Matière et mémoire* (1908), in which he describes the vital process of selecting and retaining images without recourse to the brain as a necessary condition for the image to exist. The apprehension of the image is dependent upon its existence "out there" "in the world" independent of the labor of the brain knowing it: "neither nerves nor nerve centers can . . . condition the image of the universe," he notes (1988, 19). But the brain (or brain complex), in recognizing something as an image, is engaged in the labor of "becoming": it is the active center of a vortex of fluxes and flows continually redefining the shape and tenor of the material world (139).

The implications of this antisolipsistic trope are significant inasmuch as they suggest a real ethicopolitical and social realm of objects and world. In other words, things really are, for Bergson, if they are always perceived in flux: we don't just perceive them as projections. And we can be wrong about what we perceive. It is not all in the head (or the nerve centers). What's more, we remember objects and events only inasmuch as the memory conjured is useful to the body—as agent—for performing an action (such as recognizing a significant object, acting upon an impulse, making tea, or fixing the fence). Memories are not stored in the brain ready to be retrieved any more than objects are fixed in the mind; they are instead noted as a series of collated impressions *in time*, juxtaposed onto images retrieved quickly for these purposes. This dynamic, relational model of self and other, object and memory, foregrounds contingency as well as fluidity: a throbbing network of images vitally reconfiguring reality as it shifts and changes with our ever-morphing phenomenal (and neural) selves.

What this suggests is a retort to neurological or neurobiological determinism (and Platonism) by which everything that we know and do is a function of preexisting objects or neural processes and brain function. Instead, a form of duality is preserved, acknowledging the relationship between object-in-the-world and mechanisms-for-apprehending-it, while fêting the self behind the body as a form of pure perception continually reaching into the past to refashion an always recursive present, a "pure" presence, perhaps, of being and becoming, creatively working its way through the detritus of history and past impressions to forge itself always anew. This is not so much Benjamin's Angelus Novus as a matter-spirit demon flitting back and forth between the real (or Real) and the imagined, creation and recognition, invention and perception, I and others; in other words, an ontology and epistemology of flow.

Pictures in and outside the Brain: What Neural Metaphor for the Future?

I have argued that this neural metaphor, and this neuroscientific turn, suggest as much a return to science post-postmodernism as an attempt to reconcile the modernist desire-for-science with a familiar postmodernist relativism, offered in the metaphorized tropes of the networks, pathways, and flows. In many ways these operate as objects of syncretism in the way postmodernism sought conjunctions between modes and objects of analysis. In this sense, I am taking postmodernism as an aesthetic moment tending more rather than less toward syncretism, connections, juxtapositions, and heterogeneity; and less toward purism and simplicity. This is distinct from Latour's argument that we have never (even) been modern in the sense that we have never purely believed in science over mysticism, certainty over mystery. It is rather postmodernism posited as a form of knowing emphasizing the joining of opposites and distinct regimes: the desirability of an additive and juxtapositional rather than a subtractive or purificatory mode. In short this is a post-postmodern turn, an extension of postmodernism wrapped back into science, poeticized and integrated into philosophical thought.

This may also herald a new form of attention to bodies (taken as complex networked entities made up of pathways and cells that operate at the level of exchange), and to how we operate as groups and as collectivities; how information is exchanged as a condition of being and intellection proper to the bioinformation age: an age in which information processing perhaps overdetermines our modes of self-understanding, implicitly and often explicitly conjugated with the biological (the cellular, the neural). What we find is a trend toward metaphorizing ourselves as machines and as biological systems, simultaneously: we operate like the machines (computers, networks), that is, we operate according to machinic conjunctions and disjunctions, flows. But these are also just like the biological conjunctions, disjunctions, and flows that we know through neurology. These are posited and conceived as biological processes, and, specifically, as neural ones—at every level. This is true in the strict scientific sense and at the level of the image itself, the metaphor, which helps to image—to imagine—to intellect, to conceive, to understand—a contemporary way of being in the world not divorced from nature but permeated by "it" through and through: a nature redolent of the dynamic, complex, and connected body-machine.

If, as Henri Laborit has argued, we are sad at the death of a loved one because we are literally contained in them: our memories are deposited as neural memories in their bodies and cells, conjured at a dinner party, or a family reunion; then we are inextricably tied to one another, in bodily conjunctions intensified through proximity and repetition. In this regard, the neural turn is more than "just" metaphoric: it responds to and it describes real social processes that impact on our philosophical selves at the deepest level: the level of our conceptions of the self, the other, and the sense of time passing: history, futurity, memory, and society. The productive ambivalence of the metaphoric mode both expands and transforms our self-conceptions, and thus our selves, while echoing discursive fields anchored in equally heteronomic processes of intellectual production and richly imaginal, ever-morphing modes of understanding. We are not "just" juxtaposing an image onto another, but creatively and productively coloring, twisting, refashioning the world in which we live through this imaginal node.

NOTE

1. I am indebted to Joe Kelleher for pointing me toward Virno, Žižek, and Ricœur at a vital stage in the elaboration of this argument.

WORKS CITED

Arendt, Hannah. 1969. *On Violence*. San Diego: Harcourt, Brace.
Badiou, Alain. 2005. *Being and Event*. Translated by Oliver Feltham. London: Continuum.
Bergson, Henri. 1988. *Matter and Memory*. Translated by N. M. Paul and W. S. Palmer. New York: Zone Books.
Canguilhem, Georges. 2000. "The History of the History of Science." In *A Vital Rationalist: Selected Writings from Georges Canguilhem*, edited by François Delaporte, translated by Arthur Goldhammer, 49–63. New York: Zone Books.
DeLanda, Manuel. 2006. *A New Philosophy of Society: Assemblage Theory and Social Complexity*. London: Continuum.
Deleuze, Gilles, and Félix Guattari. 1980. *Capitalisme et schizophrénie II: Mille Plateaux*. Paris: Les Éditions de minuit.
Descartes, René. 1996. *Les passions de l'âme*. Edited by Pascale D'Arcy. Paris: Flammarion.
Gamwell, Lynn, and Mark Solms. 2006. *From Neurology to Psychoanalysis: Sigmund Freud's Neurological Drawings and Diagrams of the Mind*. Binghamton, NY: Binghamton University Art Museum.
Laborit, Henri. 1976. *Éloge de la fuite*. Paris: Gallimard.
Ricœur, Paul. 2000. *La mémoire, l'histoire, l'oubli*. Paris: Éditions du Seuil.
Tarde, Gabriel de. 1993. *Les lois de l'imitation*. Edited by Bruno Karsenti. Paris: Éditions Kimé.
Virno, Paul. 2008. *Multitude: Between Innovation and Negation*. Translated by Isabella Bertoletti, James Cascaito, and Andrea Casson. New York: Semiotext(e).
Žižek, Slavoj. 2006. *The Parallax View*. Cambridge: MIT Press.

Part 2

The Neuroscientific Turn in Practice

To join two unlike things, as in the poetic process of metaphor or analogy, is to provide a special kind of insight—what rhetorician Kenneth Burke famously called "perspective by incongruity." One of the most remarkable values of transdisciplinary work, we believe, is its invocation of this perceived incongruity—bringing together vocabularies, methods, and epistemologies that might at first glance appear to be mutually exclusive—to illuminate a subject from new angles. In this part, we present a sampling of some work that can be broadly categorized as part of the turn we describe in this book. Some of the essays participate in more familiar neurodisciplines like neuroethics (Racine and Zimmerman), yet others (Hendrix and May, Birge, Gorzelsky) are undertaking research in *emergent* neurodisciplines: neurohistory, neurotheology, neuroliterary criticism, and neuroliteracy. Each essay in this part exhibits a penchant for challenging academic boundaries—an inclination characteristic of transdisciplinary work, illustrating the productive clash of neuroscience with novels (Birge), history and theology (Hendrix and May), literacy studies (Gorzelsky), and pragmatic philosophy (Racine and Zimmerman). Each essay asks us to approach its subject from a different direction, to understand and undertake research in traditional disciplines from new perspectives. The final piece in this part is a response essay written by science and technology scholar Anne Beaulieu. In responding specifically to the essays from part 2 of the collection, Beaulieu invites readers to think critically about what their home disciplines can bring to the translational research emerging from the neuroscientific turn. In addition, Beaulieu's essay frames the kinds of intellectual inquiry and intellectual generosity necessitated by scholarship that reads across disciplinary lines.

Chapter 4

Brainhood, Selfhood, or "Meat with a Point of View"

The Value of Fiction for Neuroscientific Research and Neurological Medicine

Sarah Birge

Describing the dearth of collaboration between neuroscientists and humanists who study consciousness, literary scholar Norman Holland provides a dynamic vision of the two groups' research.

> I think of neuroscience and the human sciences as like two very small human beings energetically tunneling in from opposite sides of a very large Alp. . . . I imagine the neuroscientists tunneling in from the west, and the humanists from the east. I am afraid the neuroscientists on the west side of the Alp do not listen much to sounds of digging from the humanists on the east side. Humanists, however, listen very closely to what the neuroscientists are doing; they read popular writers like Damasio, Pinker, Edelman, Ratey, Gazzaniga, and many others. (2001)

This representation of the relationship between the "two cultures" encapsulates two important observations: that both humanists and neuroscientists are seeking a common goal—an understanding of consciousness and mind—and that the current state of communication between the two fields leaves much to be desired. However, the dialogue may not be as one-sided as commonly thought.

The interchange of ideas between neuroscience and the humanities is frequently represented as something more akin to information transfer than collaboration. Scholars in almost every academic discipline have recently turned to the neurosciences for inspiration and validation, taking up contemporary neuroscience research and using it to bolster scholarship in areas that include literature, psychology, philosophy, and anthropology, among many others. This fascination with all things "neuro" extends beyond academic discourse through media representations that distill and distort the results of scientific research, fueling a cultural imaginary that is highly brain-centric. The primary feature of this new "neuroculture," as Giovanni Frazzetto and Suzanne Anker (2009, 815) have termed contemporary society, involves the attempt of scientists and science journalists to explain everything from economics to fashion to emotion through neuroscience. In our neuroculture, selfhood is replaced by "brainhood," the condition of "*being*, rather than simply *having*, a brain" (Vidal 2009, 6). Rather than identity being understood as a convergence of environmental and biological factors, the problematic notion of brainhood implies that all the components of self (such as memory and agency) are explained by neurological functions, making embodiment irrelevant for cognition and consciousness.

In addition to the attention given to the brain in the sciences and popular media, the culture of brainhood is also being propagated by a number of humanities and social science fields. Literary criticism has gained its own "neuro" following in recent years. Termed "Neuro Lit Crit" in a *New York Times* "Room for Debate" blog,[1] cognitive approaches to literature have been thriving in the past two decades, resulting in titles such as *The Neural Imagination: Aesthetic and Neuroscientific Approaches to the Arts; The Neural Sublime: Cognitive Theories and Romantic Texts; The Work of Fiction: Cognition, Culture, and Complexity;* and *Shakespeare's Brain: Reading with Cognitive Theory.* Like many "collaborations" between the humanities and sciences, these studies tend to draw research results and concepts from neuroscience, focusing in particular on the neural processes involved in our reading, writing, and understanding of stories. Or at least this is the common understanding: as exemplified in Holland's Alp metaphor, the transfer of information from science lab to literary analysis is usually assumed to be unidirectional. Although scholars of literature and science (who are most often humanists) have long extolled the value that interdisciplinary work can have for both the sciences and the humanities, they often fail to provide any indication that scientists share their views about these exciting topical conjunctions.[2]

This study seeks to address that gap, questioning how the recent spate of "neurofiction"—that is, novels that feature neuroscience in their themes and structure—is actually noticed and employed by those in neuroscientific

fields. In particular, I would like to draw attention to the work of a growing number of neurologists who believe that studying neurofiction can
improve scientists' understanding of the interactions between brain and
mind. I argue that not only are the clinical neuroscientists cited here more
aware of contemporary fiction than common "two cultures" stereotypes
would have us believe, they are using that fiction to enhance their abilities
of scientific inquiry. By questioning the relationship between applied neuroscience research and imaginative fictional representations of the mind, I
hope to demonstrate that fiction is an important tool for scientific exploration that deserves further attention from the neuroscientific field.

Recent neurofiction builds on a long tradition of exploring subjectivity
through literature (e.g., many works of modernist authors such as Proust,
James, and Woolf examine issues of consciousness). Current incarnations
of interiority in fiction incorporate and at times challenge circulating neuroscientific research, allowing us to question the construction of our brains
and selves as they are formed in social, cultural, and scientific contexts.
Rather than reducing the mind to the brain, however, these novels question the notion of "brainhood" (Vidal 2009), highlighting the impossible
divisions contemporary science attempts to draw between biology and
experience.

In this essay, I explore the provocative question of whether novels
contribute to or count as scientific research. Philosopher Daniel Dennett
thinks so, as he indicates in his praise of the fiction of Richard Powers:
"sometimes outsider observers can hit upon felicitous ways of putting
things that have escaped the searches of the primary workers, and when
this happens, it isn't just teaching. It is a contribution to the research itself"
(2008). Taking Dennett's claim as a starting point for examining the contributions of contemporary novels to neuroscience from the perspective of
neuroscientists themselves, I aim to demonstrate that neurofiction's ability
to create and explore selfhood, rather than merely brainhood, positions the
genre to fill an important role in studies of consciousness.

The Rise of the Neuronovel

Despite Dennett's support of recent novels, several literary critics see neurofiction as little more than a gimmick employed by contemporary authors
to improve their cachet in an oversaturated fiction market. Marco Roth,
editor of the literary journal *n+1*, summarizes this viewpoint in his essay
"The Rise of the Neuronovel" (2009). Roth demonstrates his distaste for
the neuroscience-themed work of current authors—such as Ian McEwan,

Jonathan Lethem, Rivka Galchen, and Mark Haddon—who he feels are merely "cribbing from contemporary case studies" in neuroscience, using explanations for action that are so biologically particular to individual brains that the story lines they produce cannot possibly resemble the lives of most readers (149). For Roth, contemporary neurofiction is yet another symptom of our reductionist neuroculture in which the brain has come to stand for the mind.

Neurofiction, according to Roth, frustrates a "readerly *meaning impulse*" by which the reader would typically generalize experiences in fiction to make them compatible with his own. He takes as his specific examples recent novels that feature a main character with a cognitive disability, such as Mark Haddon's 2003 novel *The Curious Incident of the Dog in the Night-Time* (Asperger's syndrome), Jonathan Lethem's 1999 novel *Motherless Brooklyn* (Tourette's syndrome), and John Wray's 2009 novel *Lowboy* (schizophrenia). Dismayed that the authors "load almost the entire burden of meaning and distinctiveness onto their protagonists' neurologically estranged perceptions of our world," Roth claims that the neuronovels employ "odd language for describing odd people, different in neurological kind, not just degree, from other human beings" (147). By focusing on potential neurobiological influences on behavior (which are actually infrequently mentioned in most neurofiction), Roth fails to take sufficient notice of the value of these "neurologically estranged perceptions," which represent and enact the process of self-making that occurs in every brain, disordered or not, and which has long been an object of study in both literary and scientific fields.

Contemporary neuronovels that delve into themes of mind and self do more than just mirror science or selectively sample research data. They provide meditations on consciousness, on the experience of what it is like to have a brain and a mind, and on the limits of our tools for understanding that experience. Popular science writing, such as that of Oliver Sacks or V. S. Ramachandran, often depicts cognitive disorders (by exploring the doctors' experiences with brain-damaged patients), but fiction is able to go one step further by representing the subjective viewpoint of people who might have a difficult time describing their own perspective narratively due to their cognitive deficits. Neurofiction, which frequently features central characters with cognitive disabilities, provides complex portrayals of the intersection between brains and culture, serving to elucidate ways in which the interactions of biological structures and processes, physical environments, and social interactions (including institutions such as the legal and health-care systems) operate to construct individual and social understandings of cognitive disability. Neurofiction articulates and influences the webs of meaning that contribute to the cultural creation and experience of cognition and consciousness.

Roth ignores these aspects of the neurofiction he writes about, seeming to desire a separation between biological and psychological causes for human nature and action. For example, he feels that novels such as *Lowboy* could provide useful information about the teenage experience if only the main character's actions and beliefs were not a result of cognitive disability. This view of schizophrenia in *Lowboy*, as something that is laid over (and could be separated from) the "real" story, oversimplifies the complicated process of diagnosis, the social contributions to identity, and the many historical and cultural influences on definitions of the normal. One of the more interesting things about Wray's novel is that throughout most of the story, as Lowboy (the sixteen-year-old main character) pursues his love interest, breaks rules, and traverses New York City, it is impossible to separate the teenager from the schizophrenia. A comprehensive view of Lowboy's life and thoughts complicates an overly medicalized interpretation of his actions like the one to which Roth seems to subscribe. By analyzing characters' behavior much as neurologists often view patients', out of the context of life and the many factors that influence the way we experience the world, Roth actually participates in the very reductionism that he critiques. As medical humanities scholar Martyn Evans observes, humans are "meat with a point of view"; an understanding of the experience of consciousness is just as essential to scientific study as are the neurobiological underpinnings of that experience (2003).

Consciousness in Context

Recognizing the importance of considering brains within a broader context of selfhood, a growing number of neurologists are describing the potential value of incorporating neurofiction into their research and practice. Neurofiction has been featured in a number of recent science periodicals, including peer-reviewed research articles, career development publications, and patient newsletters and magazines. For example, three publications of the American Academy of Neurology frequently review neurofiction: *Neurology* (the academy's official peer-reviewed journal for clinical neuroscience), *Neurology Today* (a biweekly newspaper for neuroscientists), and *Neurology Now* (a patient education magazine). *Advances in Clinical Neuroscience and Rehabilitation (ACNR)* includes a regular feature titled "Neurology and Literature," and books such as *The Curious Incident* and *Saturday* (McEwan 2006) have been reviewed by doctors in the *British Medical Journal* (Thistlethwaite 2003; Quin 2005). Haddon's novel is also included in a list of otherwise nonfiction "key reading" texts on Asperger's syndrome in *Advances in Psychiatric Treatment* (Berney 2004), and a quotation from the novel introduces a peer-reviewed article on cognitive processing in autism (Behrmann et al. 2006). Similarly, a quotation from *Out of Mind* (Bernlef 1989) introduces an article written to educate pharmacists on the pathology of dementia (Thomas and Rai 2001).

While these examples demonstrate that many neurologists are indeed aware of contemporary fiction on neuroscientific themes, the often perfunctory nature of their reviews (which most often eschew interpretation or evaluation in favor of summary and a general recommendation for their colleagues to read the books) suggests that the literature is not being explored to its fullest potential. Some of the reviews listed merely categorize every found instance of a particular illness in fiction (e.g., Larner 2007, 2008), while others do more disservice to the novels by performing extremely superficial readings that do not adequately address the complexity of the texts. For example, in a *Lancet* article titled "Prejudice in a Portrayal of Huntington's Disease," Nancy Wexler and Michael Rawlins recommend against *Saturday*, arguing that the character Baxter "is the worst caricature of someone with Huntington's disease. His calumny is blamed entirely on the length of the CAG repeat on his fourth chromosome, and his villainy is due only to his disease. McEwan sadly reinforces the stigma and stereotypes from which families with Huntington's disease suffer, and which make them hide both their inheritance and their destiny" (2005, 1069). While characterizations of Baxter's appearance and movements are open for critique, McEwan does not merely reduce Baxter to his Huntington's. The authors of this review do not take into consideration the fact that the entire novel is focalized not through Baxter (thereby not giving us direct access to his own assessment of his motivations) but through Henry, a neuroscientist. In fact, the novel's limited perspective on the disease underscores many of the pitfalls of science that might be avoided through further study of literature. Henry, because of the habits developed through his training, sees everything through the context of the neuro. He not only reduces Baxter's actions to neurobiological causes but also interprets his own thoughts and feelings in terms of his brain activity, demonstrating the ease with which it is possible to begin seeing brains instead of selves.

Another productive but little-explored area of study for neurologists is the construction of disease through the telling of stories. Neurofiction often involves doctors as major characters, which provides opportunities to examine the complex ethical decisions involved in "reading" and "writing" disease as doctors attempt to detect and interpret a patient's symptoms. Decisions about treatment and cure are not as clear-cut when viewed in the context of personal circumstances surrounding diagnosis and cognitive disorder as it manifests throughout a person's life. In Liam Durcan's *Garcia's Heart* (2007), for example, a neuroscientist is asked by a lawyer to present brain scans in court in order to demonstrate that a friend on trial for war crimes was not acting of his own accord. In *Saturday*, the neuroscientist protagonist Henry Perowne must decide whether or not to operate on the man who broke into Perowne's home and threatened his family. In Mark Salzman's *Lying Awake* (2000), the main character is a nun who lives for her transcendent moments of connection with God during which she produces copious spiritual writings. After learning that her experiences are caused by epilepsy, she reluctantly decides to have surgery, which leaves her lacking purpose and spiritual conviction. In each of these examples, decisions about how to interpret and treat the symptoms of potential brain dysfunction are not easily determined, as social and biological factors must be considered if patients and doctors are to make choices based on the experience of the disease rather than on medical descriptions alone.

Reality Simulators, Empathy Generators

The intimate perspective on neurological disorders that Roth identifies as a primary flaw in these novels is also of course the very feature that has caught the attention of a number of contemporary neuroscientists. Doctors and researchers in many fields of brain science promote the reading of neurofiction to accomplish three major goals: improving education on cognitive disorders (including education of doctors, patients, and the public), improving narrative or analytic skills, and heightening empathy in caregivers. In *European Neurology*, for example, Uroš Rot reviews neuroscientific themes in the novels of Ian McEwan, suggesting that neurologists take note of the novelist's accuracy and observational skill: "After all, the ability to observe and precisely document his findings are qualities of a good clinician" (2008, 15). Barbara Koppel, chief of neurology at New York Metropolitan Hospital, explains to colleagues that the book provides "the neurologist with a sneak peek into the complex lives our patients lead and the impact their condition has on their soul," an understanding of feeling to which "quality of life scales and depression inventories" cannot compare (2002, 28). Perhaps the emphasis on notions of brainhood rather than selfhood in neuroscience results from the lack of appropriate tools that

Koppel notes. An understanding of consciousness presently remains out of our grasp, and current technologies and skill sets in neuroscience cannot adequately describe the experience of having a brain. Neural scans do not depict selves.

The "sneak peek" into an individual's experience of self provided by neurofiction affords doctors an insider's view on the development of disease as it occurs through both social and biological influences on a patient's life. Neurofiction can combine multiple viewpoints and study lived experiences in ways that traditional research formats cannot. For example, typical case studies focus on describing the symptoms of one particular disorder with limited discussion of context. Literary accounts, on the other hand, can demonstrate the interactions between an individual and multiple environmental and social factors that affect the ways in which that person experiences the world. Danish Epilepsy Center neurologist Peter Wolf notes that literary accounts of epilepsy are useful for epileptologists because they reflect societal attitudes about the disorder and provide additional perspectives (e.g., from the patient's point of view) for the doctor to consider (2006, 10). This feature is also pointed out in a *Neurology Today* review of *Icy Sparks* (Rubio 1998), which tells the story of a young girl diagnosed with Tourette's syndrome. The novel is noted for its ability to provide "realistic insight into the psychosocial difficulties experienced by many affected with TS. It should be read by every physician diagnosing and treating TS patients" (Jankovic, Kadmon, and Jankovic 2001, 25).

The value of neurofiction to provide vivid representations of the mind lies partly in its ability to immerse the reader in a character's worldview. Psychologists Raymond Mar and Keith Oatley describe fiction as a model that abstracts the world. This model, they claim, functions similarly to computer simulations, allowing the user to broaden her experiences and develop skills that will be useful beyond the model. Two important simulative functions occur through reading: (1) readers experience thoughts and emotions as though they were participating in the stories, and (2) stories model and abstract social relations, "allow[ing] for prediction and explanation while revealing the underlying processes of what is being modeled" (Mar and Oatley 2008, 173). Because of novels' simulative powers, neurofiction can provide a means for engaging with minds for which first-person descriptions are unlikely. Stories rely on abstraction, a process of selection that "clarifies understandings of certain generalizable principles that underlie an important aspect of human experience, namely intended human action" (175). Literature thus provides a model of what it might be like to be someone else, synthesizing complex information into a compelling story that helps us better understand social interactions. Literary theorists similarly describe fiction's ability to evoke emotional experiences in readers. Elaine Scarry proposes that this effect is achieved through suggestions of detail that the reader then fills in to create a particular idea or vision of events (1995). This feature of novels requires active participation from the reader, promoting a subjective construction of events that can have intense emotional resonances.

The ability to reproduce a first-person perspective is perhaps novels' most highly valued quality by neurologists who urge their colleagues to study fiction, as a *Neurology Today* review of Lethem's *Motherless Brooklyn* claims.

> The book provides a realistic account of what it is like to be inside the brain of a TS individual. It describes not only the outward manifestations of the inner torture, the tics, but also the intensity of the premonitory feelings and sensations that precede each movement or sound as if the tics escape from the desperate effort by the individual to contain and control them. The straining dam eventually gives. It is exhausting, scary, edgy, and exhilarating. (Jankovic, Kadmon, and Jankovic 2001, 25)

Although London neurologists Alastair Wilkins and Simon Shorvon provide one of the few less-than-favorable reviews of fictional representations of brain disorders (which they believe are often oversimplified), they believe that neurofiction can have "a vital role in emphasizing to the clinician the true impact and experience of disease." In fact, they suggest that more neurological novels should perhaps be a part of medical curricula (2005, 2473). This inclusion would complement the growing number of medical school courses that use literature to explore issues such as medical ethics. Goals of medical humanities courses often include the development of empathy through reading literature, as we see on New York University School of Medicine's Medical Humanities website, which states that "attention to literature and the arts helps to develop and nurture skills of observation, analysis, empathy, and self-reflection—skills that are essential for humane medical care" ("Mission Statement").

Although seldom acted upon, the suggestion that studying consciousness in fiction is an important part of medical education is not a new argument. William Walton Goody, a pioneer of modern neurology, was a fervent believer in the power of Proust to help students understand the human context of brain disease, even suggesting to students that *À la Recherche du Temps Perdu* would prove more useful to them than neurology textbooks (Wilkins 2007, 29). Because of its ability to represent the "feeling" of consciousness, to invite the reader into a mind they could never know directly, neurofiction would be particularly useful for neurologists and researchers who seek to explore the conjunction of individual experience, social relationships, and biological underpinnings that contribute to the effect of self.

Novelists by Night

The impulse to create their own neurofiction reveals some scientists' desires to move beyond strict medical or scientific interpretations of patients. Neuroscientist and *Still Alice* (2009) author Lisa Genova, for example, believes that any understanding of brain disease requires a shift beyond the biological data on an individual's brain to the historical and social contexts of the person. Her debut novel tells the story of a college professor who is diagnosed with early-onset Alzheimer's disease. Genova has explained that the novel was partly inspired by her grandmother's experience with the disease, an event that she could understand neither through neuroscience nor through literature alone.

> I was fascinated. I read a lot in the scientific literature about what was going on inside her head at the molecular level. I read a lot of nonfiction written by clinicians and caregivers. But I couldn't find a satisfying answer to the question, "What does it feel like to have this?" By the time my family was caring for my grandmother, she was too far along to communicate an answer to this question. But someone in the earliest stages could. This was the seed for *Still Alice*. (Simon and Schuster, under "On Books and Writing")

Genova believes good storytelling and good science go hand in hand: "All good scientists, the ones who help the field think in an entirely new way, are storytellers," she writes. "Stories are what move us as human beings and connect us to each other, whether you're talking about physics or biology or a woman who has early onset Alzheimer's." She explains that she uses her background in neuroscience to provide people with a view into a world they would normally be unable to access (Blackburn 2009).

Like Genova, many brain scientists have expanded their knowledge of their field by becoming avid writers as well as readers of fiction; in fact, a majority of contemporary neurofiction is written by neuroscientists themselves. Rivka Galchen, author of *Atmospheric Disturbances* (2008), received her MD with specialization in psychiatry from the Mount Sinai School of Medicine. Colleen McCullough, who recently completed a novel set in a neurological research center, originally worked as a neuroscientist for over two decades. Neurologist Robert A. Burton publishes medical thrillers, Princeton professor of neuroscience Michael Graziano has published five novels, neuroscientist David Eagleman of Baylor College of Medicine is also a fiction author published in twenty-one languages, and the list goes on. Many of these authors have spoken about the relationships between reading and writing and their clinical and research skills, seeing these various investigations of the human as intertwined. Liam Durcan, a neurologist at Montreal Neurological Hospital, believes that writing has made him a better doctor. He explains that interactions with patients have taught him that "narrative can change or be changed by disease." Even more important to Durcan is reading: "Being a reader means giving yourself up to a story; it makes you a better listener, and makes you more appreciative of what is being told to you and the importance of the personal history" (Antoline 2005, 69).

Durcan's observation that narrative can change disease is a crucial one. Medical humanities courses such as narrative medicine (which trains doctors to use literary techniques to understand patient narratives) have much to offer doctors and scientists, but it is no more productive for scientists to draw only superficial interpretations from literature than it is for literary scholars to promote scientific bases for literary structures with minimal evidence. Rather than seeing their patients' narratives as texts to be read or using examples from novels to describe symptoms, neurologists would gain a more complete understanding of the limits of narrative by focusing on the many layers of construction and constant revision inherent in any narrative of experience. Patient narratives are not merely texts, nor are they surfaces upon which a doctor imprints his perspective; self-narratives are written through a mutual process of inquiry and collaboration. In order to understand selfhood rather than brainhood, researchers could study and write fictional representations of consciousness to both explore and experience the cocreation of stories and selves.

Neurofiction as Neuroscience

Jonah Lehrer's *Proust Was a Neuroscientist* has inspired dialogue on whether such artistic inquiries into the nature of consciousness can accurately be described as science. His critics generally object to discussing art as science in anything but the most metaphorical terms. In a blog post titled "Who Is a Neuroscientist?" Noah Hutton (creator of the blog *The Beautiful Brain*) cautions against believing that "artists are actually making *scientific* breakthroughs themselves."

[Neuroscientists] are investigators of the central nervous system who use the scientific methods of hypothesis, observation, and deduction to generate testable, repeatable results. They focus mostly on cells, neurotransmitters, and proteins, unveiling the mechanisms that, on a massive scale, account for our thoughts and behaviors. If an individual does those things, they are a neuroscientist. Like a neuroscientist, Proust was an investigator of the nervous system; but his tool was the written word, and his methods were subjective and introspective. He was not a neuroscientist, nor were the other household names Lehrer calls upon in his book. (2010)

Hutton's claims about neuroscience reveal a problematic bias of scientists to create an unnecessary distinction between scientific knowledge (that which can be verified through testing) and more creative forms of investigation or truth-seeking. Creative literary works, in this view, may be mined for useful information about doctors or disease, but they are generally not seen as producing new knowledge themselves. However, if the generation of testable results were removed as a requirement, Hutton's description would very much apply to novelists: they use observation and deduction to generate hypotheses, "what if" questions that seek to identify the source of our thoughts and behaviors.

The product of these methods of inquiry and innovation—that is, the novel—might best be viewed as a technology for self-exploration rather than a scientific text that describes the brain. Affective registers and imaginative reconstructions are as essential to understanding the experience of consciousness as identifications of neural correlates. In an interview for the Harvard Book Review, novelist Richard Powers comments on the overlapping speculative goals of science and art.

All arts and all sciences are components of that speculative, recursive, self-revising process of making and testing meaning. . . . To tell stories— from simple fables to complex political sagas—is to speculate and test speculations about the world as we find it. . . . Every child begins as a scientist. But every child's science is also high art; only adults enforce the distinction between testable theories about the world and wild stories about it. (Palay 2010)

Powers touches on a crucial aspect of literature's power: its engagement with and generation of imagination. Imagination is not often described as a fundamental tool for scientists, but without the ability to generate new ideas and picture things not as they are, but as they might be, scientists would be hard pressed to continually produce new and surprising results. As McEwan notes, the work of the novelist is as essential for exploring human life as that of the scientist because "novelists can go to places that might be parallel to a scientific investigation, and can never really be replaced by it: the investigation into our natures; our condition; what we're like in specific circumstances" (Tonkin 2007). Rather than try to define novels as science experiments, then, we might more productively focus on what each technology can bring to our current exploration of the human condition.

NOTES

1. "Can 'Neuro Lit Crit' Save the Humanities?" 2010. *Room for Debate* (blog), *the New York Times*, April 5, 2010. http://roomfordebate.blogs.nytimes .com/2010/04/05/can-neuro-lit-crit-save-the-humanities/.

2. For examples of works by humanists that describe literature's influence on science, see "Science Fiction at the Bench" (Milburn 2010), *Chaos and Order: Complex Dynamics in Literature and Science* (Hayles 1991), *Literature and Science as Modes of Expression* (Amrine 1989), and *Membranes: Metaphors of Invasion in Nineteenth-Century Literature, Science, and Politics* (Otis 1999). Although scholars generally avoid citing the originality of ideas in one field over the other, some recent books such as *Different Engines: How Science Drives Fiction and Fiction Drives Science* (Brake and Hook 2008) and *Proust Was a Neuroscientist* (Lehrer 2007) attempt to establish a more predictive or causal role for literature's impact on scientific disciplines. These connections are tenuous at best, though they do describe interesting thematic overlaps between ideas generated in literature and in science.

WORKS CITED

Abi-Rached, Joelle, and Nikolas Rose. 2010. "The Birth of the Neuromolecular Gaze." *History of the Human Sciences* 23, no. 1: 11–36.

Alcauskas, Megan, and Rita Charon. 2008. "Making a Case for Narrative Medicine in Neurology." *Neurology* 70:891–94.

Amrine, Frederick, ed. 1989. *Literature and Science as Modes of Expression*. Dordrecht: Kluwer.

Antoline, Dawn. 2003. "Up Close and Personal with John H. Menkes, MD: Pediatric Neurologist, Novelist, and Playwright." *Neurology Today* 3, no. 12: 34–35.

Antoline, Dawn. 2005. "A Focus on People: An Interview with Liam Durcan, MD: Neurologist and Short Story Writer." *Neurology Today* 5, no. 2: 69.

"Author Revealed: Lisa Genova." Simon and Schuster website. http://authors .simonandschuster.com/Lisa-Genova/49420182/author_revealed.

Behrmann, Marlene, Galia Avidan, Grace Lee Leonard, Rutie Kimchi, Beatriz Luna, Kate Humphreys, and Nancy Minshew. 2006. "Configural Processing in Autism and its Relationship to Face Processing." *Neuropsychologia* 44, no. 1: 110–29.

Berney, Tom. 2004. "Asperger Syndrome from Childhood into Adulthood." *Advances in Psychiatric Treatment* 10:341–51.

Bernlef, J. [Hendrik Jan Marsman]. 1989. *Out of Mind*. Translated by Adrienne Dixon. Boston: David R. Godine.

Blackburn, Maria. 2009. "Still Lisa." *Bates Magazine Online*. http://www.bates.edu /x205051.xml.

Brake, Mark L., and Neil Hook. 2008. *Different Engines: How Science Drives Fiction and Fiction Drives Science*. London: Macmillan.

Brust, John. 2001. "'The Diagnosis' Offers Horrific Message for Neurologists." *Neurology Today* 1, no. 3: 26.

Chung, Sean. 2009. "Review of *Still Alice*." *Neurology Now* 5, no. 4: 13.

Crane, Mary Thomas. 2000. *Shakespeare's Brain: Reading with Cognitive Theory*. Princeton: Princeton University Press.

Dennett, Daniel C. 2008. "Astride the Two Cultures: A Letter to Richard Powers, Updated." In *Intersections: Essays on Richard Powers*, edited by Stephen J. Burn and Peter Dempsey, 151–61. Champaign, IL: Dalkey Archive Press.

Dresser, Rebecca, and Peter J. Whitehouse. 1994. "The Incompetent Patient on the Slippery Slope." *Hastings Center Report* 24, no. 4: 6–12.

Durcan, Liam. 2007. *Garcia's Heart*. New York: Thomas Dunne Books.

Eco, Umberto. 2006. *The Mysterious Flame of Queen Loana*. Translated by Geoffrey Brock. Orlando, FL: Harvest Books.

Evans, Martyn. 2003. "Roles for Literature in Medical Education." *Advances in Psychiatric Treatment* 9:380–86.

Frazzetto, Giovanni, and Suzanne Anker. 2009. "Neuroculture." *Nature Reviews Neuroscience* 10:815.

Galchen, Rivka. 2008. *Atmospheric Disturbances*. New York: Farrar, Straus and Giroux.

Genova, Lisa. 2009. *Still Alice*. New York: Simon and Schuster.

Haddon, Mark. 2003. *The Curious Incident of the Dog in the Night-Time*. New York: Doubleday.

Hayles, N. Katherine, ed. 1991. *Chaos and Order: Complex Dynamics in Literature and Science*. Chicago: University of Chicago Press.

Holland, Norman. 2001. "The Neurosciences and the Arts." *PSYART: A Hyperlink Journal for the Psychological Study of the Arts*. http://www.psyartjournal.com /article/show/n_holland-the_neurosciences_and_the_arts.

Hutton, Noah. 2010. "Who Is a Neuroscientist?" *The Beautiful Brain: An Online Magazine*. http://thebeautifulbrain.com/2010/03/who-is-a-neuroscientist/.

Jankovic, Joseph, Sheryl Kadmon, and Cathy Jankovic. 2001. "Two Novels Address Challenges of Living with Tourette Syndrome." *Neurology Today* 1, no. 4: 25.

Kaplan, Peter W. 2004. "Mind, Brain, Body, and Soul: A Review of the Electrophysiological Undercurrents for Dr. Frankenstein." *Journal of Clinical Neurophysiology* 21, no. 4: 301–4.

Koppel, Barbara. 2002. "Epilepsy as Religious Construct." *Neurology Today* 2, no. 3: 28.

Larner, Andrew. 2007. "'Neurological Literature': Epilepsy." *Advances in Clinical Neuroscience and Rehabilitation* 7, no. 3: 16.

Larner, Andrew. 2008. "'Neurological Literature': Cognitive Disorders." *Advances in Clinical Neuroscience and Rehabilitation* 8, no. 2: 20.

Lehrer, Jonah. 2007. *Proust Was a Neuroscientist.* New York: Houghton Books.

Lethem, Jonathan. 1999. *Motherless Brooklyn.* New York: Doubleday.

Lightman, Alan. 2000. *The Diagnosis.* New York: Pantheon.

Mar, Raymond, Maja Djikic, and Keith Oatley. 2008. "Effects of Reading on Knowledge, Social Abilities, and Selfhood." In *Directions in Empirical Studies in Literature: In Honor of Willie van Peer,* edited by S. Zyngier, M. Bortolussi, A. Chesnokova, and J. Auracher, 127–37. Amsterdam: Benjamins.

Mar, Raymond, and Keith Oatley. 2008. "The Function of Fiction Is the Abstraction and Simulation of Social Experience." *Perspectives on Psychological Science* 3:173–92.

Massey, Irving. 2009. *The Neural Imagination: Aesthetic and Neuroscientific Approaches to the Arts.* Austin: University of Texas Press.

McEwan, Ian. 2006. *Saturday.* London: Vintage.

Miksanek, Tony. 2001. "Review of *The Diagnosis,* by Alan Lightman." *Journal of the American Medical Association* 285, no. 8: 1073–74.

Milburn, Colin. 2010. "Science Fiction at the Bench." *Isis* 101:560–69.

"Mission Statement." New York University School of Medicine Medical Humanities website. http://medhum.med.nyu.edu/.

Otis, Laura. 1999. *Membranes: Metaphors of Invasion in Nineteenth-Century Literature, Science, and Politics.* Baltimore: Johns Hopkins University Press.

Palay, Adam. 2010. "An Interview with Richard Powers." *Harvard Book Review,* May 2. http://www.hcs.harvard.edu/~hbr/main/current-issue/adam-palay-an-interview-with-richard-powers.

Purtilo, Ruth B., and H. ten Have. 2004. *Ethical Foundations of Palliative Care for Alzheimer Disease.* Baltimore: Johns Hopkins University Press.

Quin, John. 2005. "Review of *Saturday,* by Ian McEwan." *British Medical Journal* 12, no. 330: 368.

Richardson, Alan. 2010. *The Neural Sublime: Cognitive Theories and Romantic Texts.* Baltimore: Johns Hopkins University Press.

Richardson, Alan, and Ellen Spolsky, eds. 2004. *The Work of Fiction: Cognition, Culture, and Complexity.* Aldershot, UK: Ashgate.

Rot, Uroš. 2008. "Ian McEwan—Novels about Neurological and Psychiatric Patients." *European Neurology* 60, no. 1: 12–15.

Roth, Marco. 2009. "The Rise of the Neuronovel." *n+1*: 139–51.

Rubio, Gwyn Hyman. 1998. *Icy Sparks.* New York: Viking.

Rushton, Simon K., and Rob Gray. 2006. "Hoyle's Observations Were Right on the Ball." *Nature* 443, no. 7011: 506.

Salzman, Mark. 2000. *Lying Awake.* New York: Knopf.

Scarry, Elaine. 1995. "On Vivacity: The Difference between Daydreaming and Imagining-Under-Authorial-Instruction." *Representations* 52:1–26.

Shem, Samuel. 2002. "Fiction as Resistance." *Annals of Internal Medicine* 137, no. 11: 934-37.

Singh, Varun, and Andrew Larner. 2009. "Some More Dickensian Diagnoses." *Advances in Clinical Neuroscience and Rehabilitation* 9, no. 2: 19–20.

Thistlethwaite, Jill. 2003. "Review of *The Curious Incident*, by Mark Haddon." *British Medical Journal* 327:815.

Thomas, H., and G. Rai. 2001. "The Aetiology and Pathology of Dementia." *Hospital Pharmacist* 8:33.

Tonkin, Boyd. 2007. "Ian McEwan: I Hang on to Hope in a Tide of Fear." *The Independent*, April 6. http://www.independent.co.uk/arts-entertainment/books/features/ian-mcewan-i-hang-on-to-hope-in-a-tide-of-fear-443501.html. Accessed June 7, 2010.

Vidal, Fernando. 2009. "Brainhood, Anthropological Figure of Modernity." *History of the Human Sciences* 22, no. 1: 5–36.

Wexler, Nancy S., and Michael D. Rawlins. 2005. "Prejudice in a Portrayal of Huntington's Disease." *The Lancet* 366, no. 9491: 1069–70.

Wilkins, Alastair. 2007. "How Proust Can Make You a Better Neurologist." *Advances in Clinical Neuroscience and Rehabilitation* 7, no. 1: 28–29.

Wilkins, Alastair, and Simon Shorvon. 2005. "Review of *Saturday*, by Ian McEwan, and *The Curious Incident of the Dog in the Night-Time*, by Mark Haddon." *Brain: A Journal of Neurology* 128, no. 10: 2470–73.

Wolf, Peter. 2006. "Descriptions of Clinical Semiology of Seizures in Literature." *Epileptic Disorders* 8, no. 1: 3–10.

Wray, John. 2009. *Lowboy*. New York: Farrar, Straus and Giroux.

Chapter 5

Neuroscience and the Quest for God

Scott E. Hendrix and Christopher J. May

Modern writers such as Will Durant have described the medieval period as an "age of faith" (1950, iii), and despite the danger of gross generalization, in its broad strokes it is difficult for a modern researcher to deny the characterization. In fact, for countless premodern Europeans during the Middle Ages and beyond, belief in God's divine plan served as an organizing principle through which they understood themselves and their world. Considering the power of the Catholic Church during these centuries, this theistic focus should come as no surprise, nor should it surprise us that many people during this period desired more than simply a chance to pray to God—many wanted to *know* God immediately and directly. For these impatient souls, mystical union with the divine became the goal. Not only might one hope to reap significant spiritual rewards, but the prestige a recognized mystic held provided considerable authority and power. The sociocultural ethos of the medieval period was thoroughly infused with a hegemonic discourse, that is, a set of cultural and ideological assumptions, encouraging the pursuit of mysticism.

However, while the Christian tradition provides models for this quest—perhaps the most significant being the Apostle Paul's experience on the road to Damascus and the writings in toto of the sixth-century Christian Neoplatonist Ps. Dionysius the Areopagite[1]—there was no established set of techniques designed to bridge the gap between humanity and God's divinity. Therefore, aspiring mystics experimented with different sets of what Michel Foucault would call "technologies of the self" (1988) in order to refashion themselves in ways that would make direct contact with God possible. Naturally, different technologies yielded different results. These differences can be understood through the lens of neuroscience. A neuroscientific analysis of the practices of mystics provides insights into the psychophysiological effects that these technologies of the self might be expected to produce. In conjunction with a historical analysis of the motivations for pursuing contact with the divine, a neurohistorical analysis provides a fuller framework for understanding medieval mystic behavior.

Neurohistory is a young approach, combining the analytical tools of the neuroscientist with those of the historian in order to provide fuller insights into the motivations, desires, and actions of historical actors than either discipline can provide individually. The basis of analysis is the presupposition that psychological factors must be understood within their sociohistorical contexts, but these factors are informed by standardized neurological structures.[2] Rather than remaining limited to the writings of these individuals, or those that were written about them, a neurohistorical approach allows the researcher to make reasonable judgments about how important figures were able to effectively exercise agency in pursuit of mystical contact with the divine. In short, neuroscience provides an added level of analysis to be coupled with postmodern techniques of culturally dependent textual interpretation. Historians have made great strides in gaining an increasingly sophisticated understanding of peoples of the past who lived very different lives from the researcher and explained their own experiences through radically different discursive formations—yet were dependent upon the same[3] neurological structures for forming thoughts and approaching the world as we are today.

Neurohistory is hinged upon an understanding of non–culturally specific aspects of our basic humanity that, when considered in conjunction with the societies and cultures that people build slowly over time, allow us to draw closer to an understanding of those people who make history than is otherwise possible. In addition, while it is widely understood in the neurosciences that experiences and behavior unfold from brain activity embedded in and dynamically interacting with multiple contexts (e.g., physical and sociocultural), a deeper understanding of the influences of those contexts is difficult to unpack with the tools of neuroscience alone. This is perhaps most evident in prevailing tendencies to interpret mystical experiences through a lens of pathology (for examples, see Saver and Rabin 1997; Vaitl et al. 2005). While some experiences may indeed issue from neuropsychiatric conditions (e.g., epilepsy), the relatively new field of contemplative neuroscience, which examines brain changes associated with different meditative practices, has revealed just how much meditation can produce profound, and typically desirable, neural and psychological changes in nonpathological populations (for overviews, see Ekman et al.

2005; Lutz et al. 2008; Shapiro and Wallace 2007). These changes are only meaningful, however, in the larger personal and sociocultural contexts in which they occur. Thus, both neuroscience and history stand to gain from joint neurohistorical analyses.

The current study will focus on two rather different people, Bernard of Clairvaux (1091–1153) and Teresa of Ávila (1515–82). Bernard was the son of a Burgundian nobleman and a deeply pious mother who attempted to live a life as close to the ascetic ideal as possible while acting as a married mother of seven (William of St.-Thierry 1147). Teresa, on the other hand, was the daughter of a minor Spanish knight whose paternal grandfather had converted from Judaism only to later face persecution at the hands of the Spanish Inquisition for allegedly continuing to favor the religion of his birth (Williams 2001). Bernard and Teresa were similar in one important way: both became recognized as successful in their quest for union with God, which in turn brought prestige and power to each of them. Yet even in this fact they were very different, for the techniques they used to gain this union were widely divergent, and if we want to understand not only these two but also the many other mystics who have played such a key role in the history of Christianity then we need to move beyond a historical approach to the level of neurohistorical analysis. For the modern researcher seeking to understand the way meditative practices allow an individual to reshape him- or herself, the study of historical actors who applied these techniques provides an untapped evidentiary base upon which to build a stronger understanding of the power of meditation. For the historian who wants to understand how mystics were able to (re)make themselves—transcending normal human limitations to such an extent that, in a Christian milieu, the mystic could plausibly claim contact with the divine—research in contemplative neuroscience provides a heuristic for a different understanding of ostensibly similar practices by historical figures. In other words, neurohistory allows the researcher to consider alternate routes into the subjectivities of historical figures. In virtue of both their similarities and differences, Bernard of Clairvaux and Teresa of Ávila function as good case studies to demonstrate the usefulness of neurohistory for bringing fresh insights to the study of historical actors.

Let us begin with Bernard: as a young boy he grew up in a household in which his mother, Aleth, was the most important influence. According to William of St.-Thierry, Bernard's friend and biographer, Aleth ate abstemiously, spent many long hours in prayer, and "ruled her household in the fear of God . . . and brought up her sons in obedience" (William of St.-Thierry 1147, 237B–238A). Her influence on the household is clear; eventually all of her children—and even her husband—would take monastic vows. It is also clear that she saw Bernard as special from an early age, even if we discount the formulaic tale that a holy man predicted great things for him while still in the womb. As a boy his parents sent him alone of all the children to Chatillon-sur-Seine to be educated by the secular canons of the cathedral of Saint-Vorles. Although Bernard had an early interest in classical Latin literature and poetry that would mark his writing for the rest of his life, spending his formative teenage years in such a religious setting would have only reinforced the deeply pious example set by his mother. It is likely that Bernard's habit of hagiographical reading developed here, as in his adult life many remarked upon the time he spent reading saints' lives. The most popular work in this genre was the *Life of St. Anthony,* the desert hermit who died in 356/7, portraying a man who lived a life of ascetic heroic self-denial in hopes of getting closer to God. Living in seclusion for much of his life, eating only wild foods—and of that only very little— Anthony nonetheless came to be seen as the founder of monasticism and attained such prestige in his own day that the Roman emperor Constantine (d. 337) and two of his sons corresponded with the eremite in search of advice (Latourette 1953).

All of this—his mother's early influence, his education at the cathedral school, and his devotional reading—would have created a set of cultural assumptions about sanctity through which Bernard would have been conditioned to see asceticism, complete devotion to God, and striving for divine union as the most praiseworthy life possible, leading to both spiritual rewards and temporal influence. In fact, judging from Bernard's later writing, he saw it as the essence of the human condition to "always be journeying back, always striving to know God" (Evans 2000, 23). This attitude seems to have begun to express itself soon after the death of his mother when he was nineteen, coming to the fore as the primary motivating influence in his life four years later in 1113 when Bernard and thirty companions, including two of his own brothers, entered the French monastery at Cîteaux.

Bernard left us with no shortage of written material from which we may understand his approach to his religious vocation. In his sermon on Psalm 138.8 he asked "Where shall I turn, so that I may turn to thee, my Lord God?" (Evans 2000, 22–23). Proceeding from that point, he deconstructs the concept of "turning," in the end stating that if he wishes to be one with God—which is his most fervent desire—then he must remake himself, turning into a different person, changing himself into a "little one" so as to learn meekness and humility. When compared to his *Sermones in*

Quadragesima de psalmo "Qui habitat,"[4] his meaning is clear. There he states, "A merely bodily and therefore outward conversion is a form and not the truth. We are told to deny our bodies their sensual pleasures, but that is not enough," before going on to clarify that neither is it enough to undergo only a spiritual change, for such a change can only occur after a bodily "conversion." What is the point of this "turning"? No less than ecstatic union with God, which though only momentary while in this life is nevertheless the most important goal toward which anyone can ever strive—at least while in *this* life. Bernard was successful in his pursuit, coming to be seen as a mystic during his own lifetime, and the rewards were indeed great; his contemporaries held his moral authority in such high esteem that kings and popes wrote seeking advice, while he directly intervened in the crisis in the church that left two men vying for the role of pope in 1130. Bernard has been called without exaggeration one of the most powerful men in twelfth-century Europe (Harris 2001).

Similarly, Teresa of Ávila's psyche developed within an environment in which the concept of self-abnegation and the quest for God was continuously presented to her not only as the most laudatory but also as the most fully actualized life possible. Sixteenth-century Spain was a culture in which no deviation from Catholic orthodoxy was permitted, as the Spanish Inquisition alertly policed the populace of "New Christians," Jews who had converted rather than face expulsion in 1492, for any hint of the existence of the *Maranos* who secretly practiced Judaism (Roth 1992). Teresa's paternal grandfather was one of these New Christians who knew all too well the suspicious nature of the inquisitors: forced to do public penance in 1485 for alleged Jewish activities, he was lucky to escape with his life (Williams 2001). Even among the larger body of those with no links to Judaism the authorities were extremely vigilant, especially for signs of Erasmianism, which had become popular among members of the intelligentsia, such as the former chancellor of Alcalá University, Pedro de Lerma (d. 1541), imprisoned in 1537 for his Erasmian teachings (Kamen 1997). This humanistic reevaluation of Catholic doctrines based upon the writings of Desiderius Erasmus (d. 1536) was far too liberal for the deeply conservative brand of Catholicism practiced on the Iberian Peninsula in the sixteenth century.

It is important that we have some sense of the extent to which religious orthodoxy informed the time and place of Teresa's birth, for as with Bernard, the result was the creation of a hegemonic discourse built around cultural assumptions that would have provided her with a view of her relationship to God and her faith that would have been absorbed from the culture in which she lived. True, there are elements of her writings that indicate a carefully negotiated resistance to the overarching power structures of her society,[5] but negotiation is not rejection. Her religious identity would have been reinforced by the education she received from the Augustinian nuns at Ávila and the letters of St. Jerome (d. 430) that she read assiduously. The impact of these letters on Teresa cannot be underestimated, as they present a form of Christian asceticism and complete devotion to God that clearly influenced Teresa in her own pursuit of union with the divine; in her autobiography she not only regularly refers to Jerome's letters but also indicates that others compared her to the fifth-century saint (Teresa of Ávila 1565).

Teresa had the potential to gain even more from recognition as a mystic than had Bernard. First of all, as a woman her freedom of action and influence was extremely circumscribed in sixteenth-century Spain.[6] Through recognition as a mystic, though, she was able to carve for herself a sphere of action, becoming a reformer and founder of monasteries. By the time of her death some thirty-four monastic establishments, split equally between those housing monks and those housing nuns, owed either their existence or their reform and reestablishment to her guiding hand. Certainly she faced opposition, as many within the Inquisition suspected her of heresy and diabolical discourse, but ultimately she prevailed through the direct support of King Philip II (d. 1598) and Pope Gregory XIII (d. 1585). It is unthinkable that she would have been able to wield such power had she not been widely seen as regularly engaged in direct discourse with God (Mujica 2003).

We can understand both Bernard and Teresa better if we approach them as people who employed what Michel Foucault would call "technologies of the self," if we understand these technologies to comprise both inward mental techniques as well as those directed at disciplining the body with the intended goal of changing the inner landscape of the individual (Foucault 1988). This application of the theory is permissible because Foucault's essential insight was that there is a venerable Western tradition against allowing one's "self" to represent a passive construct, coming about through an interaction of biological and environmental conditions. Rather, from at least the time of the Greeks there have been many people who have taken a proactive role in the process of self-determination and individuation, using processes such as self-reflection—either carried out through a process of individual meditation or through analytical writing about one's innermost thoughts and feelings—and confession based on guided self-analysis of what faults a person might have and how the person confessing would like to change who he or she is. In this way an individual (re)constructs the self as a subject, manipulating and remaking individual characteristics.

It is essential to recognize, though, that this process does not occur in a vacuum. The techniques employed, as well as the desired ends, are shaped both by the society within which one lives and by the hegemonic discourse dominant in that society. This discourse informs individuals about what is expected of them and what, precisely, people should strive to be, a concept perhaps most fully developed by the early twentieth-century theorist Antonio Gramsci (Gramsci 2007; Martin 2002). Hegemonic discourse is one of the key components in the socioculturally constructed self, wherein researchers seeking to understand historical actors should take account of the intersecting elements that inform the construction of the self for each individual. Within high and late medieval society, the biographies of the Desert Fathers (with Anthony being the most prominent), the writings of church fathers (such as Jerome), and the work of Ps. Dionysius all served as constitutive elements of this discourse in which mystical contact with God was widely held to be the most desirable—and profitable—thing one could accomplish in this life, as detailed above. It is within this orienting framework that Bernard and Teresa operated. However, while the ends to which Bernard and Teresa devoted themselves were strongly determined by the prevailing hegemonic discourse, the methods for attaining contact with the divine were not. In the absence of any sort of guiding principles, the technologies applied by would-be mystics were necessarily idiosyncratic and ad hoc.

Turning first to Bernard, while he employed many technologies of the self in an ascetic project of self-fashioning, for the purposes of this study the most important component of this program was a twofold strategy designed to remake himself as one capable of having direct experience of the divine. First, from the time that Bernard entered into his novitiate at the monastery of Cîteaux, he habitually avoided sleep, with his contemporaries noting that he stayed awake in prayer long after the other monks retired for the evening or went back to bed after rising to perform one of the nightly offices (William of St.-Thierry 1147). While research indicates that people can get by on very little sleep—if the sleep received is regular—details of Bernard's life suggest that he chose random intervals at which to maintain long vigils of prayer, indicating that he would have experienced not only sleep restriction but also sleep fragmentation, a combination that would have caused a destabilization of the waking state (Doran et al. 2001; Durmer 2005).

It is normal that during sleep, activity in the prefrontal cortex of the brain diminishes substantially, most strikingly in an area known as the dorsolateral prefrontal cortex (DLPFC) (Muzur et al. 2002). Typically, this activity recovers quickly upon waking, but in a person enduring sleep deprivation and fragmentation, recovery can be delayed. The DLPFC is central to the so-called executive functions, including working memory, reasoning, temporal ordering of information, directed thought, and "reality monitoring." Therefore, an individual undergoing extended sleep fragmentation—as was likely the case with Bernard—would experience numerous brief periods in which the likelihood of abnormal perceptions and the inability to properly categorize these experiences as such are greatly increased (Wesensten et al. 1999). It is important to note here that while a neuroscience perspective provides insights into the likely effects of Bernard's practices, it makes no epistemic or ontological claims about those experiences.

Bernard's intentional sleep deprivation and fragmentation increased the likelihood of having experiences that might be classified as mystical. To the extent that sleep deprivation and fragmentation were employed as a means for remaking the self—as it appeared to be for Bernard—then there were valuable technologies of the self. In addition, Bernard practiced long periods of intense prayer typically directed toward a single object, often standing for such extended periods as he did so that his feet would swell and his knees would buckle from the effort (William of St.-Thierry 1146). Prayer has not always been fully appreciated as a technology usefully employed by mystics, likely because the vision of prayer in the popular imagination is that of an undirected exercise of limited length requiring little if any concentration, useful perhaps for reducing "mental clutter" but for little else (Kroll and Bachrach 2005). However, the way in which Bernard used prayer—performed for extended periods while concentrating on a single object—means that we should think of what he was doing more as meditation than any simplistic appeal to the divine. Meditation allows one skilled in its use—and we must remember that Bernard practiced intensive prayer for decades—to experience alternate states "by promoting more frequent transitional periods [between sleep and waking], and by training the meditator to remain experientially aware for longer intervals" during transitional periods (Austin 1998, 464). This process of lengthening the periods in which one is experientially aware is consistent with the increase in involuntary microsleeps that follow sleep deprivation, resulting in frequent hypnagogic and hypnopomic perceptual distortions (Durmer 2005; Austin 1998). In a phenomenological investigation of intense meditation practice, Jack Kornfield reports experiences from practitioners including "still objects moving," "LSD melting-like visions," "a spider as big as my hand com[ing] out of the floor," "vision[s] of Buddha," as well as intense emotional experiences (1979, 44). Therefore, with Bernard's penchant for combining sleep deprivation with long periods of intense meditative prayer, it is small wonder that he had affective mystical experiences, such as the direct vision of the Virgin Mary and the infant Jesus at Bethlehem he experienced either during or immediately after involuntarily falling asleep while in the chapel at Vorles, as well as apophatic experiences in which he felt himself to be in direct contact with the divine (William of St.-Thierry 1147, 229A). Both of these experiences are common to those who practice intense meditation, with or without the addition of ascetic practices (Kornfield 1979). In fact, in a recent ethnographic study of "new paradigm" Christian churches that focus on cultivating a personal experience with God, Tanya Luhrmann reports that nearly half of those interviewed had an experience of God that could be classified by outside observers as hallucinatory—but is believed by the individual experiencing the event to be true contact with the divine (Luhrmann 2004).

Compared to Bernard, Teresa of Ávila's technologies of the self appear both tamer as well as closer to the traditional paradigm of a meditative practitioner intent upon obtaining a transcendent experience. In stark contrast to Bernard of Clairvaux, who practiced sleep and food deprivation among other practices, Teresa explicitly deemphasized asceticism—although she did so while providing a nod toward the usefulness of ascetic practices for the benefit of the male authority figures who read and critiqued her work (Perez-Romero 1996). To understand this last point one must know that in order to be recognized as a mystic who spoke to God rather than a heretic who spoke to demons there was an unwritten set of criteria one would typically have to meet in late medieval and early modern Spanish culture, one of which was that such experiences should be difficult to attain as well as confined to a select group of ascetics (Perez-Romero 1996). It is for this reason that in her autobiography Teresa mentions the arduous asceticism that she and her nuns follow, yet the most "arduous" practice that she specifically identifies is adhering to a vegetarian diet (Teresa of Ávila 1565). This approach is certainly a far cry from Bernard's sustained mortification of the flesh.

Furthermore, the nature of Teresa's mystical experiences was quite different than Bernard's, being entirely apophatic in nature. In fact, she rejected experiences including sensible components as false and deceptive. Instead her experiences of the divine were utterly ineffable and completely interior, described as moments of absolute stillness, confidence, and an inability to perceive the passage of time or even the existence of the self as "the soul is made one with God" (Teresa of Ávila 1577, 132). Such experiences sound so odd and inexplicable to modern readers that modern scholars have struggled to explain and define her mystical experiences. How can one reconcile periods of completely interiorized absorption that remove one from the phenomenological world that is the ordinary field of maneuver and action for grounded modern people? Such confusion has led some to seek explanation in medical conditions such as epilepsy (Barton 1982). Unfortunately such an approach is reductionist, relying too much on modern medical understanding divorced from historical context and the full range of evidence provided not only in Teresa's copious writings but also in statements her contemporaries made about her. Biological considerations are an important component of understanding mysticism, but this form of analysis must be combined with a close reading of the historical context of the subject under scrutiny. With that in mind, it is important to note that Teresa's contemporaries perceived her experiences as useful and laudatory—if highly uncommon—rather than aberrant and possibly pathological.

Furthermore, Teresa's experiences closely mirror those of others wielding similar technologies of the self. For example, in a comparative analysis, Ken Wilber[7] likens Teresa's attainments to a stage called "conditional nirvikalpa samadhi" in Eastern contemplative traditions (Wilber 2000a, 645). This replicability of the effects of meditative technologies speaks against a pathology-interpretation of Teresa's experiences.

The question becomes, then, how did she accomplish her goal? What she needed was a system of meditation, allowing her to transcend her self—that essential construct created, in part, by the hegemonic discourse within which she existed, which produced a set of unspoken assumptions that made up how she thought about herself and her world (cf. Bourdieu 1998). In the absence of an established system of meditation within sixteenth-century Catholicism, she invented one. Foucault's understanding of the technologies of the self closely mirrors Teresa's system, beginning as it does with an intense process of self-reflection during which time the individual categorizes and reflects upon the sins weighing down the soul in order to gain entry to the first mansion—in Teresa's terminology—proceeding through a process of a "gradual letting go of 'unnecessary things,'" thereby allowing one to rise through the higher mansions by way of a process of self-reflective analytical prayer (Ahlgren 1979). The progressively unfolding fruits of her efforts are described in her *Interior Castles* as seven mansions. Each mansion was phenomenologically unique and, she believed, one step further on her path to transcendent union with God (Teresa of Ávila 1577). Ultimately, as described in Ken Wilber's analysis of this work, "The soul is 'so completely absorbed and the understanding so completely transported' . . . that the soul seems incapable of grasping anything that does not awaken the will to love" (2000a, 305). It is through this internal absorption, occurring at mansions four through seven in Teresa's system, with longer duration at each higher level, that she comes into contact with God.

Teresa's descriptions of her absorptive states are important, as they parallel descriptions provided by others in similar states. Indeed, J. H. Austin has studied internal absorption in relation to Zen practitioners and notes six regular characteristics of internal absorption: (1) no spontaneous thought; (2) an intensified, fixed, internalized awareness; (3) an expansion of especially clear awareness into ambient space; (4) the disappearance of the bodily self; (5) a distinctive closing off of all sight and sound; (6) a deep, blissful serenity; and (7) a marked slowing or cessation of respiration (1998).[8] With the possible exception of the final characteristic, this description seems strikingly close to the descriptions Teresa provides of her own experiences. Use of meditation to alter the effective functionality of the brain is not unknown. Recent research shows that during the meditations of long-term practitioners, many parts of the brain experience functional "deafferentation" (Newberg et al. 2001). That is, they become deprived of their normal inputs. As a result, the practitioner's phenomenological experience shifts dramatically, taking on the characteristics of absorption distilled by Austin. Austin speculates that during absorption, this functional deafferentation is effectuated by excitation of a part of the brain called the reticular nucleus of the thalamus (2006). The thalamus is considered to be the "relay station" of the brain, because signals for sight, sound, touch, and taste synapse onto the thalamus before proceeding to their primary sensory neocortical areas. Many higher-order brain areas also feed back into the thalamus. For this reason, one way to significantly impact the sensory processing of the brain is to alter the thalamus, such that outputs are no longer projected to the cortex (indeed, this is partly what happens in deep sleep). Excitation of the thalamus's reticular nucleus results in the inhibition of other thalamic nuclei, which would normally take in sensory signals. As a result, Austin notes that several domains would drop out of consciousness: vision, hearing, proprioception, vestibulation, and somatosensation. As a consequence of the loss of these signals, our normal cues to orient ourselves in space are lost, and so, conceivably, one's sense of space would be dramatically altered or even vanish entirely.

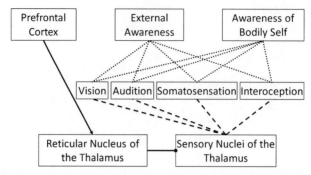

Fig. 1. Hypothesized neurophenomenological interactions producing a state of absorption. The arrowhead represents an excitatory interaction, the round-head represents an inhibitory interaction, and dashed lines represent the dropping out of phenomenal awareness.

The initial excitation of the reticular nucleus would come from projections from the prefrontal cortex (Zikopoulos and Barbas 2007), a key area subserving the practitioner's ability to maintain attention on an object of meditation. Long-term meditation is associated with increases in prefrontal cortical thickness (which may be likened to the increased capacity of a larger muscle; Lazar et al. 2005). Thus, practitioners such as Teresa would be more capable of cutting off the flow of information from the thalamus, thereby producing a state of absorption (see fig. 1). Through intensive meditation, then, Teresa effectively applied a technology of the self, in this case a sophisticated set of meditative techniques of her own devising, in order to remake herself into an individual capable of having experiences that, within her cultural milieu, would have been interpreted as mystical contact with the divine.

Bernard and Teresa represent two very different approaches to pursuing contact with God. Bernard's approach was marked by severe asceticism, while Teresa's by a more classical meditative system. Bernard's asceticism produced visions of the divine, which were likely mediated by the effects of sleep fragmentation on the prefrontal cortex. Teresa's meditations resulted in periods of absorption, which were likely mediated by the inhibition of the thalamus, a gateway for sensory information to reach the rest of the brain. Despite these differences, the end result was the same for both Bernard and Teresa: each attained the status of a mystic—and eventually sainthood—providing immense authority and respect, allowing for a wider field of action than either would have enjoyed otherwise.

NOTES

1. The literature on both of these figures is vast. For an introduction to Paul's influence on the development of mysticism, see James D. G. Dunn, *The Theology of Paul the Apostle* (Grand Rapids, MI: Wm. B. Eerdmans, 1998), 390–412. For the major issues relating to Ps. Dionysius, see John P. H. Clark, "The *Cloud of Unknowing*," in *An Introduction to the Christian Mystics of Europe*, ed. Paul Szarmach (Albany: State University of New York Press, 1984), 273–92; Eric D. Perl, *Theophany: The Neoplatonic Philosophy of Dionysius the Areopagite* (Albany: State University of New York Press, 2007), chapter 1, "Beyond Being and Intelligibility."

2. For a variation on this approach, see Jerome Kroll and Bernard S. Bachrach, *The Mystic Mind* (New York: Routledge, 2005).

3. There are certainly individual differences between people (accounting for certain personality differences, for example). However, it is generally accepted that Homo sapiens brains have remained architecturally stable for over ten thousand years.

4. Quoted in Evans 2000, 23, from *Sermones in Quadragesima de psalmo "Qui habitat*," 2.1, 2.4.

5. Not all would agree with my interpretation. See Antonio Perez-Romero, *Subversion and Liberation in the Writings of St. Teresa of Avila* (Atlanta: Rodopi, 1996), 85.

6. There have been a number of studies dealing with the way women at various levels of early modern Spanish society managed to negotiate an active role for themselves in a heavily patriarchal society. For examples, see Allyson M. Poska, *Women and Authority in Early Modern Spain: The Peasants of Galicia* (Oxford: Oxford University Press, 2005), 11–14, 186–93. For Teresa specifically, see Barbara Mujica, "Skepticism and Mysticism in Early Modern Spain: The Combative Stance of Teresa de Avila," in *Women in the Discourse of Early Modern Spain*, ed. Joan F. Cammarata (Gainesville: University of Florida Press, 2003), 54–76.

7. Ken Wilber is a well-known author who has published over a dozen books distilling and synthesizing Eastern and Western psychology and philosophy. While we believe Wilber's analysis is valuable here, some caution is recommended since Wilber is not an academic. However, he has published his models of consciousness in peer-reviewed venues (see Wilber 1997, 2000).

8. Austin's books are extensively referenced monographs on the neurobiology of meditation.

WORKS CITED

Ahlgren, Gillian T. W. 1979. *Entering Teresa of Avila's Interior Castle.* Translated by Kieran Kavanaugh, OCD, and Otilio Rodriguez, OCD. New York: Paulist Press.

Austin, James H. 1998. *Zen and the Brain.* Cambridge: MIT Press.

Austin, James H. 2006. *Zen-brain Reflections.* Cambridge: MIT Press.

Barton, Marcella Biro. 1982. "Saint Teresa of Avila. Did She Have Epilepsy?" *Catholic Historical Review* 68:581–98.

Bourdieu, Pierre. 1998. *Practical Reason: On the Theory of Action.* Stanford: Stanford University Press.

Doran, S. M., H. P. A. Van Dongen, and D. F. Dinges. 2001. "Sustained Attention Performance during Sleep Deprivation." *Archives of Italian Biology* 139:253–67.

Durant, Will. 1950. *The Age of Faith: A History of Medieval Civilization—Christian, Islamic, and Judaic—from Constantine to Dante: A.D. 325–1300.* New York: Simon and Schuster.

Durmer, J. S. 2005. "Neurocognitive Consequences of Sleep Deprivation." *Seminars in Neurology* 25, no. 1: 117–29.

Ekman, P., R. Davidson, M. Ricard, and A. Wallace. 2005. "Buddhist and Psychological Perspectives on Emotions and Well Being." *Current Directions in Psychological Science* 14, no. 2: 59–63.

Evans, G. R. 2000. *Bernard of Clairvaux.* New York: Oxford University Press.

Foucault, Michel. 1988. "Technologies of the Self." In *Technologies of the Self: A Seminar with Michel Foucault,* edited by Luther H. Martin, Huck Gutman, and Patrick H. Hutton, 16–49. Ann Arbor: University of Michigan Press.

Gramsci, Antonio. 2007. *Prison Notebooks.* Translated and edited by Joseph A. Buttigieg. New York: Columbia University Press.

Harris, John Glyndwr. 2001. *Christian Theology: The Spiritual Tradition.* Portland: Sussex Academic Press.

Kamen, Henry. 1997. *The Spanish Inquisition.* New Haven: Yale University Press.

Kornfield, Jack. 1979. "Intensive Insight Meditation: A Phenomenological Study." *Journal of Transpersonal Psychology* 11:41–58.

Kroll, Jerome, and Bernard Bachrach. 2005. *The Mystic Mind: The Psychology of Medieval Mystics and Ascetics.* New York: Routledge.

Latourette, Kenneth Scott. 1953. *A History of Christianity, Beginnings to 1500.* San Francisco: HarperCollins.

Lazar, S. W., C. E. Kerr, R. H. Wasserman, J. R. Gray, D. N. Greve, M. T. Treadway, M. McGarvey, B. T. Quinn, J. A. Dusek, H. Benson, S. L. Rauch, C. I. Moore, and B. Fischle. 2005. "Meditation Experience Is Associated with Increased Cortical Thickness." *Neuroreport* 16, no. 17: 1893–97.

Luhrmann, Tanya. 2004. "Metakinesis." *American Anthropologist* 106:518–27.

Lutz, A., H. A. Slagter, J. D. Dunne, and R. J. Davidson. 2008. "Attention Regulation and Monitoring in Meditation." *Trends in Cognitive Sciences* 12, no. 4: 163–69.

Martin, James. 2002. "Between Ethics and Politics: Gramsci's Theory of Intellectuals." In *Antonio Gramsci: Critical Assessments of Leading Political Philosophers,* edited by James Martin. London: Routledge.

Mujica, Barbara. 2003. "Skepticism and Mysticism in Early Modern Spain: The Combative Stance of Teresa de Avila." In *Women in the Discourse of Early Modern Spain,* edited by Joan F. Cammarata. Gainesville: University of Florida Press.

Muzur, A., E. F. Pace-Schott, and J. A. Hobson. 2002. "The Prefrontal Cortex in Sleep." *Trends in Cognitive Sciences* 6:163–68.

Newberg, V, A. Alavi, M. Baime, M. Pourdehnad, J. Santanna, and E. d'Aquili. 2001. "The Measurement of Regional Cerebral Blood Flow during the Complex Cognitive Task of Meditation: A Preliminary SPECT Study." *Psychiatry Research: Neuroimaging Section* 106:113–22.

Perez-Romero, Antonio. 1996. *Subversion and Liberation in the Writings of St. Teresa of Avila.* Atlanta: Rodopi.

Roth, Cecil. 1992. *A History of the Marranos.* 5th ed. New York: Sepher-Harmon Press.

Saver, J. L., and J. Rabin. 1997. "The Neural Substrates of Religious Experience." *Journal of Neuropsychiatry and Clinical Neurosciences* 9:498–510.

Teresa of Ávila. 1565. Trans. and pub. 2008. *The Book of My Life.* Translated by Tessa Bielecki and Mirabai Starr. Boston: New Seeds Books.

Teresa of Ávila. 1577. Trans. and pub. 2008. *The Interior Castle.* Translated by E. Allison Peers. Radford: Wilder Publications.

Vaitl, D., J. Gruzelier, G. A. Jamieson, D. Lehmann, U. Ott, G. Sammer, U. Strehl, N. Birbaumer, B. Kotchoubey, A. Kubler, W. H. R. Miltner, P. Putz, I. Strauch, J. Wackermann, and T. Weiss. 2005. "Psychobiology of Altered States of Consciousness." *Psychological Bulletin* 131, no. 1: 98–127.

Wallace, B. A., and S. L. Shapiro. 2006. "Mental Balance and Well-being: Building Bridges between Buddhism and Western Psychology." *American Psychologist* 61, no. 7: 690–701.

Wesensten, N. J., T. J. Balkin, and Gregory Belenky. 1999. "Does Sleep Fragmentation Impact Recuperation?" *Journal of Sleep Research* 8:237–45.

Wilber, Ken. 1997. "An Integral Theory of Consciousness." *Journal of Consciousness Studies* 4, no. 1: 71–92.

Wilber, Ken. 2000a. *Sex, Ecology, Spirituality: The Spirit of Evolution.* 2nd ed. Boston: Shambhala.

Wilber, Ken. 2000b. "Waves, Streams, States, and Self: Further Considerations for an Integral Theory of Consciousness." *Journal of Consciousness Studies* 7, no. 11–12: 145–76.

William of St.-Thierry. 1147; printed 1860. *Vita prima Bernardi.* In *Patrologia Cursus Completus Series Latina,* edited by J. P. Migne, vol. 185, 225A–268B. Paris: Garnier.

Williams, Rowan. 2001. *Teresa of Avila.* London: Continuum Books.

Zikopoulos, B., and H. Barbas. 2007. "Circuits for Multisensory Integration and Attentional Modulation through the Prefrontal Cortex and Thalamic Reticular Nucleus in Primates." *Reviews in the Neurosciences* 18, no. 6: 417–38.

Chapter 6

Literacy in a Bicultural World

Integrating Sociocultural Studies of Literacy and Neuroscientific Research

Gwen Gorzelsky

In "Brains, Maps, and the New Territory of Psychology," Anne Beaulieu persuasively argues that brain mapping "recasts [social and environmental] aspects in biological terms" (2003, 563).[1] Beaulieu draws on critiques of popular discussions of brain imaging such as Joseph Dumit's work, and on the responses of many neuroscientists to Posner and Raichle's book on the promise of brain imaging, which targets both interested lay readers and scientists. The journal *Behavioral and Brain Sciences* devotes nearly sixty pages to a précis of the book, twenty-seven peer commentaries, and the authors' responses to those commentaries (Posner and Raichle 1995). Like Dumit's text, some of the commentaries critique Posner and Raichle's work, noting that such complex, culturally determined phenomena as racism are discussed in terms of brain biology in popular texts on neuroimaging (Beaulieu 2003, 563). Beaulieu argues that these discussions redefine such phenomena as biomedical conditions, opening them to attempts at genetic and pharmacological intervention (564).[2] She also warns readers of a shift away from behavioral studies of the processes of mind over time toward a focus on anatomical measures of space in the brain (562). These critiques raise important social concerns that demand greater involvement by scholars across a range of disciplines, as Dumit (2004) also argues.

However, in a 2010 article in the *Journal of Cognitive Neuroscience*, Weber and Thompson-Schill stress the particular usefulness of neuroimaging techniques for studying behavior and behavioral treatments. They argue that despite the critiques of many skeptical neuroscientists, neuro-imaging "is not merely correlational" but can "provide causal informa-tion, although not causal certainty, about the influence of brain activity on behavior. Moreover, the influence on brain activity of experimentally manipulated variables, such as stimuli and task instructions, is unarguably causal" (2415).[3] Examples of research that integrates behavioral and neu-roimaging techniques include studies of depression that highlight the use of neural networks in developing clinical treatments (Cicchetti and Pos-ner 2005, 572). Further, Posner and DiGirolamo argue that neuroimaging "has begun to allow scientists to consider possible neural mechanisms for many of the kinds of changes involved in learning and education" (2000, 878). Posner makes comparable claims in a 2003 Opinion in *TRENDS in Cognitive Science* (452). Thus while neuroimaging has problematically increased the biomedicalization of personality traits and behaviors and to some extent supplanted behavioral studies, it has at the same time extended some strands of behavioral research and opened the door to another road as well: that of using cultural work, from therapeutic to educational to reli-gious practices, to intervene in biology at the individual and social levels.[4]

In the West, cultural work takes place in significant part through what literacy researchers call *literate practices*. Scholars define *literate practices* as reading, writing, discussion, and the interpretation and use of written texts.[5] Such uses involve understanding or addressing circumstances outside the text. For instance, literate practices include using written descriptions of medical conditions to assess one's health, reading news articles on finance to make investment decisions, and using spiritual texts to manage a per-sonal crisis. These practices both reproduce and revise cultural norms (Gee 2008, esp. 103–14; Graff 2001). For the past thirty years, literacy scholars in disciplines ranging from anthropology, education, history, and linguis-tics to psychology and composition/rhetoric have worked within the socio-cultural paradigm, generating a wealth of studies concerning the shape and function of various literate practices in diverse local contexts. While some quantitative research exists, most of this work has used qualitative, textual analytic, and/or historical methods to produce a rich understanding of how literate practices shape culture, social relations, and individuals' participa-tion in such systems and thus their felt senses of identity.[6]

I argue that cross-disciplinary research studies that combine the quali-tative and textual-analytic methods of the sociocultural paradigm with neuroimaging and related methods can reveal how we already use culture to intervene in biology through literate practices, as well as how we might do so more reflectively and ethically. Further, I contend that integrating these social sciences and humanities methods into neuroimaging studies can help shift recent work on neuroplasticity away from the biomedicalization of personality traits and toward a richer, more ethical understanding of behavioral treatments and educational approaches as cultural interventions in human biology. Because these interventions occur at the nexus of culture and biology, they should be studied—and developed—in cross-disciplinary research teams that include humanists and social scientists, as well as life scientists. Such cross-disciplinary teams have the best chance of (1) identifying appropriate behaviors and categories for investigation; (2) developing interventions responsive to ethical, educational, social, and political concerns; and (3) analyzing the implications of existing and new interventions with respect to such concerns.

I work from an understanding of literate practices as inherently biocultural phenomena. David Morris defines the *biocultural* as the cultural environments and practices that emerge from "interactions between culture and biology" (Morris 2006, 541). To make the case that literate practices are biocultural, I turn first to sociolinguist James Gee's work on the acquisition of new literate practices, a process that involves apprenticeship with experienced users. I argue that understanding this process as a form of neuroplasticity offers important insight into the role of cultural practices, particularly educational practices broadly defined, in reshaping human biology. To document an example, I turn next to three meta-analyses of neurological research on the effects of meditation on brain biology. I argue that this body of neurological research underscores the potential for using cultural practices to intervene in human biology. In doing so, it implies the need for cross-disciplinary research on the effects of such intervention—that is, the biological, experiential, social, and political effects, understood as working in dialogue with one another.

At the same time, existing neurological research only begins to consider the possibilities for learning how meditative practices may revise brain biology. I outline the more detailed understanding researchers may gain. I do so by using textual analysis to identify the complex, intertwined literate practices elicited by the guided meditations authored by Thich Nhat Hanh, a Vietnamese Zen teacher influential in Europe and the United States. I suggest directions for future studies that would combine neuroimaging with textual and qualitative analyses. I argue that textual analysis is needed to identify literate practices for investigation, while qualitative research is needed to document shifts in research participants' felt experiences, relationships, and interaction in social networks. Such qualitative data is required to contextualize the neurological changes that occur and to consider their ethical, social, and political implications.[7]

Literacy Acquisition: A Biocultural Process

Gee's influential sociolinguistics approach defines literacy as proficiency in a discourse, a way of thinking, seeing, feeling, doing, and being in the world. Gee holds that literacy often draws substantially on print texts but encompasses a far broader array of behaviors. To explain how proficiency develops, Gee makes a vital distinction between acquisition and learning. Learning results from explicit instruction, for instance, in the kinds of rhetorical moves and self-presentation likely to help a speaker or writer establish credibility in her target discourse. As his examples suggest, such moves and self-presentation differ significantly depending on whether the discourse involved is that used by lawyers and judges in a court case or that used in a multiethnic urban schoolyard. Explicit instruction and the learning it generates are crucial to developing proficiency in a new discourse.[8] Yet, as Gee shows, acquisition is equally vital, and perhaps more so. Acquisition involves practicing aspects of the discourse, such as the rhetorical moves and self-presentation required to project a persuasive ethos. To succeed, such practice requires scaffolding, or supportive feedback, encouragement, and correction from proficient users of the Discourse. In short, achieving proficiency in a new Discourse requires apprenticeship in its literate practices (Gee 2008, 44–45, 48–49, 170–71).

Gee's model provides important insight into how literate practices revise one's experience of self. It reveals the extensive effects of developing proficiency in a new set of literate practices on this self. Because such proficiency requires learning to see and think in the terms of the new practices, for instance, to think like a lawyer or to perceive the world as a streetwise teen, developing proficiency reshapes perception, cognition, and even values and emotional responses. Such phenomena are all, in significant part, biological. As recent work on neuroplasticity shows, their biological dimensions are (re)shaped by one's life experience, that is, by culture (Damasio 2003, 55–57). Thus Gee's model implies that culture (re)shapes individuals' biology through the means of literate practices. Understanding *how* literate practices accomplish this work requires investigating the kinds of neuroplasticity that result from acquiring new forms of such practices. Analyzing acquisition as a process that mediates neuroplasticity provides an important window into how training in new literate practices may reshape brain biology. In the next section, I present an example of how the cultural practice of meditation does such work and then turn to the literate practices embedded in some forms of meditation.

Meditation: A Biocultural Practice

In a meta-analysis of research on the neurological effects of mindfulness meditation, Chiesa, Brambilla, and Serretti conclude that existing research demonstrates that mindfulness meditation prompts neuroplastic changes in the brain. In defining the category "mindfulness meditation," they include Zen, Vipassana, and therapeutic practices like Mindfulness Based Stress Reduction (MBSR) and Mindfulness Based Cognitive Therapy (MBCT). They compare the neural correlates of these meditation practices with those of psychotherapy, pharmacotherapy, and the placebo effect, and find significant commonalities in the cerebral areas activated by all four interventions. The major difference they note is that while antidepressants act through a bottom-up process that regulates the activity of the amygdala itself (a region that processes fear and other negative emotions), mindfulness practice, psychotherapy, and the placebo effect operate through a top-down process in which regions associated with intention and logic, like the prefrontal cortex and orbitofrontal cortex, inhibit activity in the amygdala. In his meta-analysis of research on the neurological effects of meditation practices, Austin similarly contends that meditating can decondition fear (2009, 234). It does so, Chiesa, Brambilla, and Serretti suggest, by allowing "a more flexible emotional regulation and a higher ability to detach from negative states by engaging frontal cortical structures to dampen automatic amygdala activation" (2010, 113).

Based on their review of existing research, they suggest that meditation may develop this capacity in frontal brain structures by increasing one's ability to intervene in spontaneous thought.

> Meditative training could foster the ability to control the automatic cascade of semantic associations triggered by a stimulus and, by extension, to voluntarily regulate the flow of spontaneous mentation. This result is particularly important if one considers that Farb et al. (41) reported that non-meditators showed a significant activation of left language areas in the PFC [prefrontal cortex], even when they tried to focus their attention on present experience, suggesting the impossibility to dampen their spontaneous mentation. (Chiesa, Brambilla, and Serretti 2010, 109–10)

Other recent studies confirm that meditation does decrease spontaneous thought, particularly thought associated with ongoing narrative construction of one's sense of self (Wu et al. 2010).

The importance of this capacity for intervening in spontaneous mentation grows clearer with an understanding of the function of such thought. Austin shows that the high level of baseline activity in the "resting" medial prefrontal cortex correlates with constructing a narrativized sense of self (Austin 2009, 71). In a third meta-analysis, Tirch similarly explains that "the narrative mode of self-reference has been found to be correlated with the medial prefrontal cortex (mPFC), which is involved in maintaining a sense of self across time, comparing one's traits to those of others, and the maintenance of self-knowledge" (2010, 119). Experienced meditators, he says, showed weaker links between areas of the mPFC associated with narratives of self and brain regions that may play a role in translating visceral emotions into consciously recognized feelings. This decreased linkage "may reflect a cultivated capacity to disengage the habitual connection between an identified sense of self across time and the processing of emotional memories," which in turn may increase one's capacity to more effectively manage strong emotions (Tirch 2010, 120).

Chiesa, Brambilla, and Serretti suggest that this ability to intervene in automatic mentation and manage emotions could be quite useful in treating "psychiatric disorders characterised by excessive ruminations," including anxiety, alcohol addiction, and obsessive-compulsive disorder. Some studies, they note, suggest that mindfulness meditation could help treat "all such disorders" (2010, 110). They explain that both anxious and depressive disorders may result from (1) excessive, unregulated amygdala activity; (2) the failure to effectively manage amygdala activity through subregions of the PFC, including the mPFC; or (3) both (110). They contend that the findings they review contribute to building "a neurobiological basis for the usefulness of MM [mindfulness meditation] in several clinical conditions that are only partial[ly] responsive to current treatments options[,] such as anxiety and mood disorders" (112). Austin similarly holds that approaches like mindfulness-based stress reduction "have much to recommend them as parts of a conservative approach to treating less severe forms of PTSD [post-traumatic stress disorder]" (2001, 234–35).

Tirch also advocates the use of meditation in treating this set of disorders, while his review emphasizes the potential of meditation to cultivate compassion for oneself and others. Compassion-Focused Therapy (CFT) presumes that generating compassion contributes to the capacity to regulate strong emotions. Citing studies that correlate "higher levels of reported self-compassion . . . with lower levels of depression and anxiety,"

Tirch also highlights research providing "the first structural evidence of 'experience-dependent cortical plasticity' associated with meditation training" (2010, 115, 118). As Tirch explains, Lazar et al. demonstrated a thickening in the middle prefrontal area and the right insula, a thickening that increased with more meditation experience. The middle prefrontal area is linked to caregiving and compassion, while the insula helps manage communication between the middle prefrontal area and limbic, or emotion-processing, regions. In addition, Tirch cites a 2006 review of EEG and fMRI imaging studies of meditation that indicates increased blood flow to the anterior cingulate cortex (ACC) and other areas. The ACC is associated with both the ability to focus one's attention and, along with the mPFC, attachment and caregiving. It is believed to play a role in emotion regulation and conflict resolution as well. Working from these hypotheses about its functions, Compassion-Focused Therapy attempts to "activate specific brain systems such as those linked to empathy (e.g., the insula, ACC) in order to facilitate new ways of processing difficult emotional memories or emotions" (Tirch 2010, 119).

Thus while research on the neurological effects of meditation is not yet conclusive, it provides definitive evidence of structural changes in some areas of the brain and of increased neurological activity in other areas. These changes indicate clearly the role of this cultural practice in reshaping the brain's biology. The neurological, experiential, and social significance of these changes is still under investigation. I contend that such investigations should include research on the role of specific aspects of meditative activity in fostering the kinds of neurological changes described above. Particular types of literate practices elicited by guided meditations seem well designed to promote specific neurological changes, and their neurological effects therefore should be investigated in tandem with research on the experiential, social, and cultural implications of such neurological changes.

Three questions that emerge from the meta-analyses just discussed involve the role of such literate practices in fostering these changes. First, do some literate practices elicited by guided meditations help train meditators to regulate the flow of spontaneous mentation by decreasing activity in the mPFC? Second, do other literate practices thicken the prefrontal area and insula and so augment the capacity of the PFC to regulate the amygdala and other emotion-processing regions? Third, do yet other literate practices change the ACC in ways that help cultivate the capacity for compassion for self and others, as well as additional capacity for emotional regulation? In all cases, researchers must also investigate what changes occur in research subjects' experiences and relationships in correlation with such neurological changes.

In the next section, I present an overview of my analysis of sixty-six guided meditations to show how such analysis can identify specific literate practices that should be investigated in studies pursuing the questions above.[9] Further, I argue for incorporating qualitative research into studies of the neurological effects of the acquisition of key literate practices. Through qualitative investigation, researchers will develop a deeper understanding of the experiential, social, and cultural implications of neurological changes in context of research subjects' lives and relationships, both personal and systemic.

Literate Practices in Guided Meditations: A Biocultural Intervention

My analysis includes the guided meditations in two of Nhat Hanh's books that, together, offer a fairly comprehensive collection of his work in this genre (Nhat Hanh 1987, 1993). Using detailed written instructions, guided meditations lead meditators through approaches to dealing with circumstances outside the text, ranging from their own emotions, thoughts, and perceptions to relationships with others to major life challenges, from job loss to serious illness. Meditators use these instructions to recognize and modify their physiological, psychological, and mental states; develop sustained visualizations; address strong emotions; understand motivations; revise their desires and attachments; and move from conceptual to holistic thinking. In the longer study, I categorized all sixty-six exercises according to their primary purposes, which are enacted primarily through the literate practices they elicit from meditators. Further, I considered the functions and relationships among their salient features.[10] Here, I describe the categories, their relationships, and questions they raise for neurological research.

The literate practices embedded in category 1 exercises help practitioners to cultivate mental and emotional stability, peace, joy, and attention to the present moment.[11] They do so to foster equilibrium and to help meditators better address external situations. Nhat Hanh attributes this equilibrium to "right mindfulness . . . the fountain of stability and calm that resides in each of us" (1993, 30). Like Zen teachers generally, he describes mindfulness as a change agent that allows meditators "to light up the recesses of our mind, or to look deep into the heart of things in order to see their true nature" (vii). The category 1 literate practices seem designed to train meditators to regulate spontaneous thought and the construction of narratives of self. Thus researchers should investigate whether acquiring these literate practices correlates with decreased activity in the mPFC.

The category 2 literate practices guide practitioners in accepting negative emotions, mental states, and behaviors to produce insights that transform these negative states into positive ones. Nhat Hanh stresses the destructive potential of emotions like anger, explaining that mindfulness can change their energies into more fruitful forms, such as insight and compassion. These literate practices seem calculated to help meditators regulate emotions. Thus researchers should investigate whether acquiring them correlates with (1) thickening of the prefrontal area and insula and (2) augmented capacity of the PFC to regulate the amygdala and other emotion-processing regions.

The category 3 literate practices help practitioners move from conceptual to experiential understanding of fundamental Buddhist notions like emptiness, interdependence, impermanence, and nonattachment. They aid meditators in seeing into the "true nature of phenomena," which leads to an understanding of the fundamental interconnectedness of all beings and thus to compassion (Nhat Hanh 1993, 92). They seem tailored to cultivate compassion. Thus researchers should investigate whether acquiring them correlates with changes in the ACC and in meditators' capacity for empathy, emotion regulation, and caregiving.

Drawing on the skills built by acquiring the literate practices in the first three categories, the category 4 practices help practitioners cultivate an experiential understanding of others' suffering. They promote compassion by helping meditators focus on the vulnerability of even those who have caused suffering and on the systemic roots of their actions. Like the category 3 literate practices, the category 4 practices seem designed to cultivate compassion. Therefore, again, researchers should investigate whether acquiring them correlates with changes in the ACC and in meditators' ability to regulate their emotions, empathize, and care for others and themselves.

Category 1, which helps practitioners cultivate stability and attention to the present moment, has the most exercises (thirty of sixty-six total). Acquiring its literate practices may provide a foundation for acquiring those in the other three categories. The category 2 literate practices, which help practitioners transform negative emotional, mental, and behavioral states, ask the learner to undertake efforts that seem likely to require at least some of the stability cultivated by category 1 practices. The exercises in this category form 27 percent of the total (eighteen of sixty-six), making category 2 third of four in number of exercises. Still, given that this fraction exceeds a quarter of the total, the skills these practices support form an important goal of the guided meditations. As meditators develop proficiency in category 2 practices, it seems likely that they increase their stability and attention to the present moment. These proportions suggest that the neurological changes encouraged by the category 1 literate practices may support behavioral changes needed to acquire the category 2 practices. That is, they may construct a neurological foundation for behavioral changes that in turn may promote (1) thickening of the prefrontal areas and insula and (2) augmented capacity of the PFC to regulate the amygdala and other emotion-processing regions. Thus researchers should investigate whether decreases in mPFC activity and in spontaneous thought support meditators' capacity to acquire category 2 literate practices and the neurological changes those practices seem likely to prompt.

Similar investigations are needed to understand the relationships between the category 3 and 4 practices and between those practices and the category 1 and 2 practices. Zen teachers, including Nhat Hanh, repeatedly stress the importance of shifting from conceptual to holistic knowing, a shift the category 3 practices promote. With twenty-seven of sixty-six total exercises, category 3 has the second-highest total, proportions that suggest its importance. It may lay the groundwork for acquiring the category 4 practices, which cultivate an experiential understanding of others' suffering. The small number of exercises in category 4 (four of sixty-six) may suggest the need for substantial proficiency in the category 1 to 3 practices and perhaps the extent to which acquiring those practices leads naturally to acquiring category 4 practices. The following questions require study: Do decreases in mPFC activity and spontaneous thought support meditators' ability to acquire category 3 and 4 literate practices? Do thickening of the prefrontal areas and insula and increased PFC capacity for regulating the amygdala and other emotion-processing regions do so? If so, acquiring category 1 and 2 practices may till the ground for behavioral shifts that change the ACC, thus cultivating compassion, empathy, and caregiving. Further, does acquiring practices from any of the later categories deepen meditators' acquisition of practices from earlier categories, thus, for instance, further decreasing mPFC activity and spontaneous thought, further thickening the prefrontal area and insula, and so on? Perhaps both the literate practices and the neurological changes they may support operate in a mutually reinforcing recursive relationship.

Conclusion: Integrating Neuroscientific and Qualitative/Textual Research

Whether or not it is recursive, the relationship between the acquisition of literate practices and neurological change is complexly layered. Acquisition unfolds over months or years, so its effects cannot be studied experimentally in a controlled laboratory setting, as can the effects of discrete behaviors. Yet literate practices that train perception, evaluation, and management of one's cognitive and emotional responses, as do those in guided meditation exercises, must affect practitioners' neurobiology. Questions regarding the social, political, and ethical effects of acquiring practices that change one's neurobiology are complicated and controversial, but essential. For instance, acquiring the literate practices described above probably effects one set of neurological changes with significant social and political implications, while acquiring those of a racial supremacist group probably effects another set, with equally significant implications.

While it is impossible to investigate these effects through controlled experimental studies, researchers can combine neurological investigation with qualitative and textual analysis to do so. To study the literate practices in guided meditations, qualitative researchers would need to begin with textual analysis to identify the practices to be investigated. Afterward, life scientists could design neurological investigations of the practices identified. Through qualitative interviews, focus groups, and observations, investigators could determine study participants' initial levels of acquisition of the literate practices in each category. They could then use a first round of neurological tests to establish baseline findings for meditators in all brain regions of interest.

As research participants continued acquiring practices over the next year, qualitative researchers would conduct additional interviews, focus groups, and observations. They would observe meditation group sessions, including discussion and meditation components, as well as study participants' interactions in work and social settings. Observing discussion components would produce data on participants' process of acquiring the relevant literate practices and, because participants regularly discuss their efforts to implement the new practices in professional and social contexts, on these efforts as well. Observing group practice of guided meditations would produce data on explicit instruction in the practices. These observations would offer important insights into the experiential, social, and cultural implications of the changes practitioners experience as they develop proficiency in the practices.

Throughout this process, scientists would ideally collect data on participants' neurological changes at regular intervals. They would study long-term changes, for instance possible thickening in the prefrontal area and insula, increases in ability to regulate the amygdala and other emotion-processing regions, and changes in the ACC. Further, neurologists should examine brain regions of interest while participants undertake literate practices from each category, examining, for instance, possible decreases in mPFC activity while participants undertake category 1 practices and comparing any such decreases across instances of data collection. Correlating any such changes with experiential and interpersonal shifts charted through qualitative data collection would offer insight into the implications of neurological changes and into the relationships between behavioral and neurological changes. If data collection were continued for several years, researchers could investigate longer-term changes in participants' neurological, experiential, and social states. If researchers shared neuroimaging data with participants throughout the study, its qualitative component would also afford insight into the social effects of neuroimaging itself.

Such a multimodal study would reveal how Nhat Hanh's recommended literate practices intervene in brain biology. It would also ground future investigations into how cultural activities, including other literate practices, do so. It would underscore the potential of behavioral treatments and educational efforts to offer a viable—perhaps more effective and less expensive—alternative to biomedical treatment for a range of conditions, from attention deficit and anxiety disorders to depression, addiction, and post-traumatic stress disorder. Finally, it would highlight the pressing ethical questions raised by seeing the acquisition of literate practices as an intervention in human biology. To make sound decisions about the ethics, and possibly the efficacy, of instruction in particular types of literate practices, researchers, educators, and policymakers need information on their intersecting neurological social, political, and personal effects.

NOTES

1. My sincere thanks to James Austin, Jenell Johnson, and Melissa Littlefield for very helpful feedback on earlier drafts of this chapter.
2. She makes a related argument in "The Brain at the End of the Rainbow: The Promises of Brain Scans in the Research Field and in the Media."
3. They hold that describing such studies as correlational rather than causal slights their contributions. While acknowledging the tendency in popular discussions to give too much credit to what neuroimaging can reveal, they caution against diminishing the potential of those methods as well, saying, "to demean any of our

methods cavalierly is to weaken the reputation of the whole field" (Weber and Thomson-Schill 2010, 2416).

4. Citing Begley's reporting of Schwartz's use of neuroimaging (both in documenting the effects of a behavioral treatment for patients with obsessive-compulsive disorder [OCD] and in encouraging patients to reinterpret the disorder), Beaulieu describes such uses of neuroimaging as promoting a narrative of self-improvement that can be tracked with brain scans (2000, 47, 50–51). In a more recent article, she and de Rijcke say that scans linking types of people (for instance, OCD patients) with types of brains initiates shifts in how we understand such conditions, noting, "These shifts may be towards a greater stigmatization of patients, but may also have positive consequences, where patients distinguish a subjective self from an objective brain that is visibly malfunctioning" (Beaulieu and de Rijke 2007, 739). Her point is key, and in this essay, I argue that an ethnographic understanding of patients' and research subjects' experiences and sociocultural interactions provides crucial context for analyzing the social effects of neuroimaging.

5. The term emerged from Heath's discussion of literacy events, in which talk focuses on understanding, responding to, and/or using texts (1983, 196–201, 386, n. 2) and from Street's understanding of literacy as involving "the social practices and conceptions of reading and writing" and his emphasis throughout his work on the mix of oral and literate practices typical of most societies (1984, 1).

6. Graff's (2001) work exemplifies historical use of quantitative methods. Gee's (2008), Heath's (1983), and Street's (1984) texts exemplify qualitative methods. Burton's (2008) and Madden's (1998) books exemplify historical use of textual analytic and qualitative methods.

7. In making this argument, I take up Beaulieu's call in "The Brain at the End of the Rainbow" for "a contextualization of scientific knowledge" (39). In this piece and in "The New Territory of Psychology" she defines this contextualization primarily in terms of including in popular accounts a more detailed, accurate presentation of the statistical computations underlying brain mapping. I agree with the need for this kind of contextualization and argue also for the contextualization of such representations of brain activity in qualitative data on research participants' experiences, relationships, and action in social networks.

8. This is true of discourses other than one's primary discourse, which one learns in one's family and home community, mostly by acquisition. Most of Gee's discussion (and mine) deals with developing proficiency in secondary discourses, those gained after one's primary discourse. Developing proficiency in a secondary discourse involves more effort and difficulty than does mastering a primary discourse.

9. I develop the full analysis in a book-length study I am now writing on the literate practices elicited by Nhat Hanh's guided meditations.

10. The longer study explains my methods for categorizing the exercises and identifying their features.

11. While I don't have space to illustrate examples of the literate practices, I will suggest them by noting that category 1 literate practices often ask meditators to visualize imagery that promotes peace and to cultivate responses of physical, mental, and emotional relaxation. In contrast, category 2 literate practices often

ask meditators to visualize circumstances that produce anger, fear, grief, or other strong emotions and to cultivate responses of physical, mental, and emotional equanimity.

WORKS CITED

Austin, James H. 2009. *Selfless Insight: Zen and the Meditative Transformations of Consciousness.* Cambridge: Massachusetts Institute of Technology Press.

Beaulieu, Anne. 2000. "The Brain at the End of the Rainbow: The Promises of Brain Scans in the Research Field and in the Media." In *Wild Science: Reading Feminism, Medicine, and the Media,* edited by Janine Marchessault and Kim Sawchuk, 39–54. London: Routledge.

Beaulieu, Anne. 2003. "Brains, Maps and the New Territory of Psychology." *Theory Psychology* 13:561–68. Accessed August 4, 2010. doi: 10.1177/09593543030134 006.

Beaulieu, Anne, and Sarah de Rijcke. 2007. Essay Review: "Taking a Good Look at Why Scientific Images Don't Speak for Themselves." *Theory & Psychology* 17:733–42. doi: 10.1177/0959354307081626.

Burton, Vicki Tolar. 2008. *Spiritual Literacy in John Wesley's Methodism: Reading, Writing, and Speaking to Believe.* Studies in Rhetoric and Religion 6. Waco, TX: Baylor University Press.

Chiesa, Alberto, Paolo Brambilla, and Alessandro Serretti. 2010. "Functional Neural Correlates of Mindfulness Meditations in Comparison with Psychotherapy, Pharmacotherapy, and Placebo Effect. Is There a Link?" *Acta Neuropsychiatrica* 22:104–17.

Cicchetti, Dante, and Michael I. Posner. 2005. Editorial: "Cognitive and Affective Neuroscience and Developmental Psychopathology." *Development and Psychopathology* 17:569–75. doi: 10.1017/S0954579405050273.

Damasio, Antonio. 2003. *Looking for Spinoza: Joy, Sorrow, and the Feeling Brain.* Orlando, FL: Harcourt.

Dumit, Joseph. 2004. *Picturing Personhood: Brain Scans and Biomedical Identity.* Princeton: Princeton University Press.

Gee, James Paul. 2008. *Social Linguistics and Literacies: Ideology in Discourses.* 3rd ed. London: Taylor and Francis.

Graff, Harvey. 2001. "The Nineteenth-Century Origins of Our Times." In *Literacy: A Critical Sourcebook,* edited by Ellen Cushman, Eugene R. Kintgen, Barry M. Kroll, and Mike Rose, 211–33. Boston: Bedford/St. Martin's. Originally published in Harvey Graff, *The Legacies of Literacy,* 340–70. Bloomington: Indiana University Press, 1987.

Heath, Shirley Brice. 1983. *Ways with Words: Language, Life, and Work in Communities and Classrooms.* Cambridge: Cambridge University Press.

Madden, Etta M. 1998. *Bodies of Life: Shaker Literature and Literacies.* Westport, CT: Greenwood.

Morris, David B. 2006. "Reading Is Always Biocultural." *New Literary History* 37:539–61.

Posner, Michael I. 2003. "Imaging a Science of Mind." *TRENDS in Cognitive Sciences* 7:450–53. doi: 10.1016/j.tics.2003.08.013.

Posner, Michael I., and Gregory J. DiGirolamo. 2000. "Cognitive Neuroscience: Origins and Promise." *Psychological Bulletin* 126:873–89.

Posner, Michael I., and Marcus E. Raichle. 1995. "Précis of *Images of Mind*" and "Open Peer Commentary." *Behavioral and Brain Sciences* 18:327–83.

Street, Brian V. 1984. *Literacy in Theory and Practice.* Cambridge Studies in Oral and Literate Culture. Cambridge: Cambridge University Press.

Thich, Nhat Hanh. 1993. *The Blooming of a Lotus: Guided Meditations for Achieving the Miracle of Mindfulness.* Translated by Annabel Laity. Boston: Beacon.

Thich, Nhat Hanh. 1987. *The Miracle of Mindfulness: An Introduction to the Practice of Meditation.* Translated by Mobi Ho. Boston: Beacon.

Tirch, Dennis D. 2010. "Mindfulness as a Context for the Cultivation of Compassion." *International Journal of Cognitive Therapy* 3:113–23.

Weber, Matthew J., and Sharon L. Thompson-Schill. 2010. "Functional Neuroimaging Can Support Causal Claims about Brain Function." *Journal of Cognitive Neuroscience* 22:2415–16.

Wu, Yanhong, Cheng Wang, Xi He, Lihua Mao, and Li Zhang. 2010. "Religious Beliefs Influence Neural Substrates of Self-Reflection in Tibetans." *SCAN* 5:324–31. doi: 10.1093/scan/nsq016.

Chapter 7

Pragmatic Neuroethics and Neuroscience's Potential to Radically Change Ethics

Eric Racine and Emma Zimmerman

Neuroscience research is increasingly informing ethics scholarship and ethics practices under the impetus of the fields of the neuroscience of ethics and of social neuroscience. This neuroturn within the field of ethics promises to enhance understanding of ourselves and of our fellow human beings. It has been argued that this knowledge will be a route to foster happiness in individual lives and the foundation of brain-based ethical norms and behaviors that will lead to greater social well-being (Changeux 1996, 1981). For example, Gazzaniga states that neuroscience will bring radical changes to ethics: "Neuroethics is more than just bioethics for the brain. . . . It is—or should be—an effort to come up with a brain-based philosophy of life" (2005, xv). However, an overly naturalistic approach to ethics, which is heavily or inappropriately informed by neuroscience, can lead to misunderstandings and, worse, ill-informed practices and policies. Some of these challenges have been discussed previously in the literature (e.g., the naturalistic fallacy, neurological determinism; Racine 2007), while others, more social in nature, have been fleshed out based on empirical research on the public understanding of neuroscience (e.g., neurorealism, neuroessentialism; Racine, Bar-Ilan, and Illes 2005). These latter challenges relate more specifically to the brain's unique status at the intersection of the humanities and the biological sciences. They point to the inextricable link between the study of brain function and concepts of personhood, responsibility, and autonomous decision making. In this contribution, we present and discuss the challenges of a neuroscience of ethics and of the potential contribution of neuroscience to an ethics open to empirical research. Based on historical and contemporary writings informed by neuroethics and pragmatism, and as a way forward in this debate on neuroscience's contribution to ethics, a Deweyan pragmatic approach is offered as a guide to a productive and ethically sensitive dialogue between ethics and neuroscience (Racine 2010). Because of its explicit commitment to empirical inquiry, including openness to the biological sciences, Dewey's pragmatism provides a unique theoretical backbone for our discussion of the use of neuroscience knowledge within ethics.

Neuroethics and the Neuroscience of Ethics

Although neuroethics is nascent as a thriving interdisciplinary endeavor, many views of the field have been put forward. Some scholars emphasize the importance of theoretical neuroethics issues raised by neuroscience (Roskies 2002), while others stress the need to address clinical neuroethics challenges of specific patient populations (Fins 2005). These differences between existing perspectives in neuroethics are important to bear in mind since not all scholars would agree that a discussion of neuroscience's impact on the nature of ethics is a legitimate object of neuroethics. For example, Farah and Wolpe write, "Unfortunately, the term is also used to refer to the neural bases of ethical thinking, a different topic" (2004, 26).

Although initially proposed in the early 1970s by Pontius (1973) and used in the context of neurological ethics in the 1980s by Cranford (1989), the modern use of the term *neuroethics* was reinvigorated and propelled by the writings of *New York Times* journalist William Safire (2002). While Safire was serving as the chair of the Dana Foundation, he was dedicated to promoting neuroscience research and neuroscience literacy. The 2002 Neuroethics: Mapping the Field Conference, held in California, was a key factor leading to today's developments and current perspectives on neuroethics. This meeting assembled dozens of experts in neuroscience, law, ethics, and other fields to discuss wide-ranging topics (e.g., the impact of neuroscience on the self; neuroscience and social policy; neuroscience and public discourse; see Marcus 2002). At that meeting Safire defined neuroethics "as a distinct portion of bioethics, which is the consideration of good and bad consequences in medical practice and biological research" and underscored that "the specific ethics of brain science hits home as research on no other organ does" (2002, 6–7). Shortly after this 2002 meeting, philosopher Adina Roskies was one of the first to define formally the field of contemporary neuroethics in a paper published in *Neuron* (Roskies 2002). Roskies argued that neuroethics is different from other areas of biomedical ethics because of "the intimate connection between our brains and our behaviors, as well as the peculiar relationship between our brains and our selves" (21). Roskies distinguished two parts of neuroethics: the "ethics of neuroscience" and the "neuroscience of ethics." The "neuroscience of ethics," according to Roskies, deals with fundamental concepts (e.g., free will, self-control, personal identity) and how neuroscience could change them based, for example, on neuroimaging research: "As we learn more about the neuroscientific basis of ethical reasoning, as well as what underlies self-representation and self-awareness, we may revise our ethical concepts" (22).

The next section gives an overview of ethical concepts as they come under the "neuroscience lens." Readers should keep in mind that although neuroscience research on moral behavior is captured under the banner of "neuroscience of ethics," the field of social neuroscience is usually defined in ways that overlap with or even include the neuroscience of ethics (Greene 2006). Most broadly, social neuroscience is defined as "an interdisciplinary, multilevel analysis to understanding social behavior and cognition" (Adolphs 2010, 752).

Ethics under the Neuroscience Lens

In the past two decades, social and affective neuroscience has flourished and its scope broadened since early descriptions of this field (Cacioppo and Berntson 1992). Claims about the practical implications of this research for ethics have become stronger. Representative of this movement, neuroscientist Michael Gazzaniga writes in *The Ethical Brain* (2005) that we need to be willing to change moral beliefs based on neuroscience evidence: "Knowing that morals are contextual and social, and based on neural mechanisms, can help us determine certain ways to deal with ethical issues" (177). He also suggests that the mandate of neuroethics is "to use our understanding that the brain reacts to things on the basis of its hard-wiring to contextualize and debate the gut instincts that serve the greatest good—or the most logical solutions—given specific contexts" (177).

This is not to say that the impetus of using brain-based knowledge to approach social questions is entirely new. For example, the physician and neuroscientist Paul MacLean, best known for his theory of the "triune brain," wrote with hope in 1967 that we should be attentive to how neuroscience sheds light on empathy and medical education. Only of late, however, has the neuroscience of ethics become a concerted and structured interdisciplinary endeavor. The following pages give a quick overview of recent work in the neuroscience of ethics and social neuroscience using examples in the areas of (1) moral reasoning and moral decision making; (2) moral development; and (3) moral status and personhood.

First Example: Moral Reasoning and Moral Decision Making

One of the possible contributions of the neuroscience of ethics is to shed light on long-standing disputes with respect to the nature of moral reasoning and moral decision making. This possible contribution is well illustrated by the work of the philosopher-neuroscientist Joshua Greene (Greene et al. 2001), in particular a paper on honest moral decision making. Using functional Magnetic Resonance Imaging (fMRI), Greene and Paxton (2009) examined a theoretical dispute in the realm of moral psychology. In this specific case, theories of moral behavior give several competing explanations for our ability to act honestly given the opportunity for dishonest gain: the "grace" hypothesis states that honest behavior is the result of the absence of temptation, and the "will" hypothesis advances that honest behavior stems from active resistance of temptation.

The experiment of Greene and Paxton was designed to give subjects an opportunity to act dishonestly by introducing the temptation of monetary gain and eliminating supervision. The first experiment performed allowed no opportunity for dishonest behavior, and the second relied on self-reporting, which introduced the opportunity to cheat. These two experiments were performed while subjects were scanned using fMRI; in the first experiment the participants were asked to record their prediction of the outcome of a randomly generated coin-flip with monetary reward for correct predictions and punishment for incorrect predictions. In the second experiment, the participants self-reported the accuracy of their predictions and so were given the opportunity to change their prediction to increase their monetary gain. Because the probability of accuracy in a random coin toss is known, individuals with dishonest reporting patterns could be detected as statistical outliers. Greene and Paxton predicted that the "will" hypothesis would be represented in the cortex by a greater cognitive load that would activate the "control network" as participants struggled against the opportunity to cheat. In contrast, the "grace" hypothesis would require no active resistance against temptation and no more cortical load than was used in the "no opportunity to behave dishonestly" scenario when the participant behaves honestly.

Results showed that for honest participants there was no activation of the "control network" in support of the "grace" hypothesis over the "will" hypothesis, which means that honest behavior is a result of absent temptation. Of course the study raises collateral questions. Does this knowledge diminish the praiseworthiness of honest behavior since the decision to act honestly took no extra "cognitive work" on the part of the honest subjects? Does honest behavior come *naturally* to some as a result of moral training or genetic disposition? Does this work bear implications for ways to promote honesty in real-world settings? And perhaps, more fundamentally, this study raises questions of whether brain-imaging methods bear epistemic privilege in questions of moral theory.

Second Example: Moral Development

A second contribution of the neuroscience of ethics is a better understanding of the nature of moral development based on social neuroscience and developmental neurobiology. A wide range of work in neuroscience could potentially inform views on moral development from a neuroscience standpoint (OECD 2002), but some of the most interesting and relevant work comes from the understanding of the transmission of social behavior through epigenetic mechanisms. In this case, epigenetics designates the study of changes in phenotypic expression triggered by various internal and external factors (e.g., social environment, chemical exposure, carcinogens) that cause the activation and inactivation of genes.

In a landmark study, Francis and colleagues (Francis et al. 1999) examined the role of epigenetics in perpetuating social behaviors. They started by observing that behaviors between mother rats naturally differ in the amount of licking, grooming, and arched-back nursing (LG-ABN) shown toward their pups (rat maternal behavior). In addition, female offspring of high LG-ABN mothers later exhibit the same behavior toward their pups. They found that in these offspring, there is a structural difference in the hypothalamic-pituitary-adrenal (HPA) axis in the brain, which diminishes the response to stressful stimuli (novelty). Maternal behavior influences the levels of gene expression in the HPA axis, and an increased glucocorticoid receptor mRNA expression in the hippocampus has an inhibitory effect and dampens corticotrophin-releasing factor (CRF) and adrenocorticotropic hormone (ACTH), all of which contribute to a less fearful behavioral response.

The experiment carried out by Meaney and colleagues established how a behavioral trait was transmitted across generations in rats through nongenomic epigenetic mechanisms. According to the authors' discussion, "In humans, social, emotional, and economic contexts influence the quality of the relationship between parent and child and can show continuity across generations. Our findings in rats may thus be relevant in understanding the importance of early intervention programs in humans" (Francis et al. 1999, 1158). The implications for human behavior of this research are appropriately modest, as the parallels in rats are limited (Hyman 2009). However, the epigenetic regulation of stress response based on childhood abuse have been studied in the postmortem tissue of suicide victims with similar results (McGowan et al. 2009). This latter study found that there were decreased levels of glucocorticoid receptor mRNA in the HPA axis of victims of child abuse, suggesting a causal relationship between this early childhood trauma and a compromised ability to deal with stressful stimuli. The implications of this area of research are likely unclear at this point since the mechanisms of epigenetics are still under intense examination. However, the perspective of neuroscience may bring an additional layer of depth and complexity in understanding social behavior and the transmission of ethical and nonethical behavior through generations.

Third Example: Moral Status and Personhood

Yet another possible contribution of neuroscience is to better define personhood in order to distinguish between persons and nonpersons. Determining personhood is often at the root of controversial bioethics topics where personhood is not easily defined or agreed upon. For example, the debates surrounding the determination of the moral status of the fetus and embryo, the diagnosis of disorders of consciousness, and the determination of death all summon particular definitions of personhood. Therefore, arguably some may well think that insights of neuroscience into the nature of persons could help solve these pervasive ethical challenges and philosophical perplexities. A review by Farah and Heberlein (2007) attempted to determine if neuroscience could shed light on controversies about personhood and the moral status of persons. Rather than considering how the defining characteristics of personhood map to cerebral function, they reviewed and discussed if neuroscience could clarify the biological basis for our cognitive hurdles in defining personhood.

Farah and Heberlein consider that recent developments in cognitive neuroscience and neuropsychology point to separate systems in the brain for processing persons as being distinct from other objects. Evidence from fMRI and lesion studies have isolated the fusiform gyrus as the brain region specialized for the recognition of faces and the tempoparietal junction as the region that is uniquely activated by human bodies and their similes (e.g., stick figures, silhouettes, cartoons). These regions are distinct from areas that process nonperson objects. Based on further examination and evidence supporting the existence of a person recognition system in the cortex, their paper concludes, "If our analysis is correct, it suggests that personhood is a kind of illusion" (Farah and Heberlein 2007, 45). This claim has significant repercussions beyond the cases of ill-defined personhood; claiming that personhood is an illusion supports a consequentialist theory of ethics that measures the value of an action primarily on its outcome, not fundamentally on an innate moral status or the rights of the person. The authors advise accordingly, "Rather than ask whether someone or something is a person, we should ask how much capacity exists for enjoying the kinds of psychological traits previously discussed (e.g., intelligence, self-awareness) and what are the consequent interests of the being" (46). The next section of this chapter identifies and discusses the challenges and opportunities that such research brings for ethics.

Pragmatism and the Neuroscience of Ethics

Although only a selection from a much broader field of inquiry, the examples reviewed above clearly illustrate the promises brought forward by neuroscience research. In spite of the enthusiasm of some neuroscientists and ethicists, several arguments and challenges are commonly raised against this area of neuroscience (e.g., neurological determinism, naturalistic fallacy, semantic dualism, biological reductionism, and neuroscience as a threat to ethics; Macintyre 1998; Changeux and Ricoeur 2000; Stent 1990). While acknowledging these criticisms, we offer a more tempered view based on a pragmatic framework (Racine 2007, 2008b, 2010). In our view, none of the arguments put forward is definitive, and they should rather be viewed as both cautionary warnings and practical guideposts for the future development of the neuroscience of ethics.

Neurological Determinism

The contribution of neuroscience research to ethics appears to jeopardize beliefs in free will and to support forms of determinism. For example, the classic electrophysiology experiments on readiness potentials by Benjamin Libet (1985) turned, according to some, free will into an undisputable illusion. The experiments have been interpreted as showing how different reactions and behaviors are based on automatic or unconscious responses. Thus, neural explanations may undermine beliefs in human agency and responsibility (Gazzaniga 2005). In the cases reviewed, the neuroscience of personhood may lead to thinking that our moral behaviors and intuitions are hardwired in ways that we cannot escape.

The form of strong determinism that underlies claims against free will and responsibility often relies on a flawed understanding of the complexity of biological systems and of the different epistemological strategies of neuroscience. This point has been well presented by the biologist and philosopher of biology Ernst Mayr. Mayr argued that the biological sciences do not comply with deterministic philosophies of science given the interaction of biological systems with their internal (physiological) and external (sociological) environments (1988). Biological sciences, while not committing to strong forms of determinism—where general laws would explain everything, and every event could be explained and predicted from antecedent causes—can be committed to softer, more realistic and adapted models of scientific explanation (Mahner and Bunge 1997).

Second, biological systems are open and dynamic. They are generally more complex than inorganic systems because of the emergent properties created by their molecular and cellular composition and organization. This is why Mayr stated that biological sciences do not yield predictive models as some of the basic physical sciences do (1988). Most biological explanations are not general laws but probabilities with several exceptions (Mayr 1985). This does not mean there is no continuity between organic and inorganic systems, but the organization of biological systems is different, and distinct emergent properties are generated by the organization of biological systems.

Most strong deterministic interpretations of neuroscience are based on metaphysical arguments seeking some provocative partial piece of neuroscience evidence in order to support an almost ideological belief in deterministic causation. Neuroscience is not committed to determinism, and we should welcome evidence helping us to better understand moral behavior and to pursue ethics in real-world settings. As Dewey remarked in *Human Nature and Conduct*, "We are told that seriously to import empirical facts into morals is equivalent to an abrogation of freedom. Facts and laws mean necessity we are told. The way to freedom is to turn our back upon them and take flight to a separate ideal realm. Even if the flight could be successfully accomplished, the efficacy of the prescription may be doubted. For we need freedom *in* and *among* actual events, not apart from them" (1922, 303).

Naturalistic Fallacy

As we saw above, some authors have speculated to some extent on how neuroscience could carry more normative implications. But does neuroscience of personhood, moral reasoning, or moral development fall prey to the naturalistic fallacy (Hume 1975 [1739])? For example, Farah and Heberlein underscore limitations in how we perceive and recognize persons, but they stop short of simply concluding that we should annihilate the concept of personhood. Francis and colleagues suggest that we need to take into account emerging neurodevelopmental science in early childhood interventions, but they are cautious in prescribing concrete real-world applications. Others, like Churchland, may have articulated stronger views, for example, that the neuroscience of ethics informs principle-based approaches like Kantianism and rather supports virtue-based approaches like that of Aristotle (Churchland 1998; Casebeer 2003).

Hume's historical identification of the naturalistic fallacy, and subsequent philosophical writings stressing the slide from an "is" to an "ought," are important. They prevent hasty reasoning that slides from matters of fact to value judgments. However, radical distinctions between "is" and "ought" also have serious problems themselves, such as precluding the existence of any concrete sources of the "ought." If one precludes a priori sources of ethical authority, then ethical judgments must be partly based on experience and induction (Callahan 1996). The contrary is incompatible with pragmatic naturalism, a scientifically informed ethics approach that strives to capture different disciplinary insights into the multiple levels of biological and social organization (Racine 2010).

Bioethics itself has generally rejected a strong is-ought divide by acknowledging the context-sensitive nature of biomedical ethics reasoning (Racine 2008b; Moreno 1999). And if the neuroscience of ethics (or other forms of empirical research) shows how moral reasoning is constrained and influenced by context and culture, then we need to take into account this knowledge to shape good and practical recommendations. In this sense, just as qualitative research has informed us on dying and end-of-life decision-making processes, neuroscience and moral psychology could help us further understand what moral reasoning and behavior are and how they are engaged in bioethics situations. To maximize its relevance and impact, however, the contribution of neuroscience will also need to interact with relevant non-biology-based research to fully describe and explain higher-level emergent properties of the brain underlying social interactions and moral behaviors. Dewey's concept of radical empiricism is relevant to invoke as it stresses how ethics is empirical in nature and is connected to other empirical disciplines like psychology and anthropology and even to physiology (1922). One caveat is that, within a Deweyan pragmatic perspective, the empirical understanding of ethics cannot be limited to neuroscience as is sometimes implicitly suggested by some tenets of the neuroscience of ethics. The process of "inquiry" must include the agent's subjective experience and the input of other disciplines (e.g., anthropology, sociology) and research approaches (e.g., qualitative research) that provide broader and grounded perspectives, especially on the social context of decision making (Moreno 1999; Racine 2007). From a pragmatic standpoint, the contribution of neuroscience is a welcome addition to the evolving pool of empirical bioethics research.

Semantic Dualism

Dualism, as Descartes historically described it, is the division between a material substance, the body, and a thinking substance, the mind (1992). This substance dualism has largely been discredited, but another form of dualism remains vibrant: *semantic* dualism. For example, the French philosopher Paul Ricoeur systematically argued, in his dialogue with neuroscientist Jean-Pierre Changeux, that brain properties are different from mind properties, precluding any sound integration of neuroscience on ethical thinking because brain properties are thus confused with mind properties (Changeux and Ricoeur 2000). Strong interpretations could thus consider that what Greene and colleagues are observing through fMRI has little or nothing to do with grace and honesty as defined in ethics literature. The same goes for personhood in the case of Farah and Heberlein, which could be interpreted as dealing with perception and recognition of *human individuals* but not of persons, which is a more complex ethical construct that does not equate simply with the notion of an individual human being (Glannon 2007).

Pragmatism takes a very different approach to this debate. It acknowledges, contrary to semantic dualism, that our understanding of certain mind properties like personhood and honesty can be enriched based on neuroscientific evidence (Bickle 1992). For example, in their paper on the concept of person, and in spite of the potential conflation between the notions of person and human individual, Farah and Heberlein do highlight problems in the use of this concept that cannot be dismissed simply by semantic quandaries. It is true enough that current neuroscience investigations of concepts like honesty or personhood may not capture the full scope of such complex ethical constructs on which there are multiple theoretical perspectives. Yet, these studies enrich evidence-based views of conceptual content and challenge philosophical assumptions that can be discussed and perpetuated without reality checks. If we better understand how our concepts of honesty and personhood work and at the very least become more attuned to how they work, we can increase our insight into the genesis of ethical problems created by their use.

It is clear that most ethical constructs are defined several ways, and it is naive to think that one can straightforwardly examine the neuronal underpinnings of such ill-defined and socially constructed objects. But the biology of a socially constructed object is still the biology of something— something that is shaped by the environment. Because it tempers reductionist impetuses, semantic dualism highlights the challenges in accounting for complex and subjective experience from third-party and intersubjective perspectives. However, to clarify the nature of these concepts and their properties, qualitative research and other forms of nonbiological inquiry are also required.

Biological Reductionism and Eliminativism

Some have voiced the fear that neuroscience will reduce ethical concepts to the point of examining only their trivial components, which are theoretically and practically irrelevant (Macintyre 1998). Others, including some philosophers and neuroscientists, have argued that typical folk-psychological explanations of ethics will not be reducible to lower-level explanations and will therefore be eliminated because of their inherent inaccuracy (Churchland 1986). None of the case examples reviewed above sided with such a strong eliminativist conclusion. Indeed, it would be counterproductive for the neuroscience of ethics to try to *explain away* the moral ideals that make our individual and collective existences better. However, from a pragmatic standpoint, it is hard to imagine that neuroscience investigation will not contribute to ethics. Neuroscience may not be the sole bearer of empirical knowledge on ethics, but acknowledging this is different from saying that, in principle, nothing can be learned because neuroscience targets lower-level properties. For example, if highly emotional ethical decisions were handled by different neuronal subsystems than less emotionally charged decisions, as suggested by Greene and colleagues (2001), then practical ethics could benefit from including further consideration of this aspect of moral decision making. The handling of case discussions by clinical ethicists could take explicitly into account the emotional state of patients, family members, and providers when initiating discussions (Racine 2008a). Such research could thus inform our views while providing ways to concretely address and intervene in processes involved in moral decision making and behavior. Promoting a form of revisionism based on the enrichment provided by multiple disciplinary perspectives on the mind-brain involved in moral reasoning and action is a more promising route (Racine 2010), instead of a form of theoretical eliminativism like that proposed by Churchland (1986, 2002). Once multiple levels of biological organization are acknowledged (Bunge 1977) and the potential contribution of different disciplinary perspectives actively sought based on this premise, the antineuroscience argument cannot hold beyond sheer dogmatism.

Neuroscientism as a Threat to Ethics

Perhaps at the root of some of the other arguments against the neuroscience of ethics is a fear that neuroscience will damage human values and beliefs (e.g., free will, honesty, personhood) (Editorial 1998). In the era of technologically driven biomedicine and following decades of increasingly bureaucratic models of organization of care, ethics has been considered a rampart against impersonal medicine and disrespect for persons. Part of the task of ethics is to create and protect academic and public spaces for dialogue to address the role of values in decisions and to help voice the concerns of persons and ensure that they are respected as persons and not simply considered as complex biological systems (Andre 2002). Counter to the goals of ethics, the perspective provided by neuroscience could reinforce the trend of seductive overly objective and technologically driven evidence-based medicine.

The risks of overly objective medicine should not be minimized. For example, neuroscience research and neurotechnological innovation are often reported in the media without attention to their limitations as well as the challenges inherent in study designs, such as the limited number of participants and other factors influencing external validity of results (Racine, Bar-Ilan, and Illes 2005, 2006). In addition, emerging popular interpretations of neuroscience take the form of neuroessentialist beliefs (essence of individual identity lies in the brain) and neurorealist beliefs (neuroimaging techniques directly reveal brain activation) that prepare the public psyche for hasty social use of results, that is, neuropolicy (Dumit 2004; Racine, Bar-Ilan, and Illes 2005). There appears to be evidence from a psychological and cognitive-science perspective that neuroscience explanations lead to neurorealism; they can give an added and illegitimate sense of objectivity to weak scientific explanations (McCabe and Castel 2008; Weisberg et al. 2008). Despite these overinterpretations, neuroscience does promise to bring new insights to ethics. As Dewey wrote, "Lack of insight always ends in despising or else unreasoned admiration. . . . What cannot be understood cannot be managed intelligently" (1922, 8). We therefore need to initiate changes in current training and research to fully capture the nuanced promises of the neuroturn.

To do this well, we need the humanities and other perspectives that encapsulate the broadest sets of views on ethics and on the impact of neuroscience on health-care experiences and medical decisions. To counter reductionist interpretations specifically, neuroethics will need to avoid disseminating forms of neuroscientism, that is, a form of scientism based on loaded interpretations of neuroscience and on beliefs in a neurotechnological fix that reduce our take on ethics, individuals, and society as a whole. Fostering interdisciplinary perspectives upstream in the training of neuroscientists and in research design as well as downstream in discussion of research are avenues to explore to enrich theoretical perspectives on moral concepts investigated by neuroscientists (Illes et al. 2010). Those involved in generating research must play an active role in ensuring careful interpretation and use of their research (Racine and Illes 2006). In addition, those who summon neuroscience evidence to support their claims about ethics must be held accountable to the highest scientific standards to avoid perpetuating consequential neuroscientism.

Conclusion: Pragmatic Neuroethics and the Neuroscientific Turn

This chapter reviewed some promises of social neuroscience and of the neuroscience of ethics. Insights into personhood, moral reasoning, and decision making could have wide-ranging implications for science and society. Some authors have suggested that neuroscience will fundamentally alter how we view ethics (Roskies 2002). Others have countered that neuroscience will not change anything in our understanding of ethics or that the neuroscience of ethics is inherently problematic because of logical fallacies or dangers posed to ethics (Macintyre 1998; Stent 1990). We argued from a pragmatic perspective that such counterarguments are not definitive and that, instead, they bring qualifications such as the need for careful interpretation of neuroscience research to avoid fueling reductionism and neuroscientism. The thinking laid out in this chapter highlights the need to integrate the neuroscience of ethics within an interdisciplinary understanding of ethics. However, from a pragmatic standpoint, ethics is inherently interdisciplinary, and disciplinary pluralism, along with other sources of pluralism, helps maintain a rich pool of moral insights and perspectives to conduct open-minded inquiry (Racine 2010). Neuroethics is clearly part of a broader neuroturn within ethics, and the pragmatic framework proposed in this chapter suggests a path that this turn can fruitfully explore to use the best knowledge generated by neuroscience to inform ethics. The existence of diverse points of views within neuroethics and within the neuroturn leads us to acknowledge the limitations of our own disciplinary habits and unreflexive ways of thinking. This, in return, means that we need to engage in creative, collective thinking to generate new consensuses and approaches to tackle unprecedented scholarly research and yield the way to evidence-based ethical and social progress.

NOTE

This chapter is based on material published in the following contributions: Eric Racine, "Which Naturalism for Bioethics? A Defense of Moderate (Pragmatic) Naturalism," *Bioethics* 22, no. 2 (2008): 92–100; Eric Racine, "Interdisciplinary Approaches for a Pragmatic Neuroethics," *American Journal of Bioethics* 8, no. 1 (2008): 52–53; Eric Racine, "Identifying Challenges and Conditions for the Use of Neuroscience in Bioethics," *American Journal of Bioethics—Neuroscience* 7, no. 1 (2007): 74–76; Eric Racine, *Pragmatic Neuroethics: Improving Treatment and Understanding of the Mind-Brain* (Cambridge: MIT Press, 2010). Support for the writing of this chapter comes from a New Investigator Award (E.R.) from the Canadian Institutes of Health Research as well as a grant (E.R.) from the Social Sciences and Humanities Research Council of Canada. The authors would like to thank the editors of this book for helpful and constructive comments on earlier versions of this chapter.

WORKS CITED

Adolphs, Ralph. 2010. "Conceptual Challenges and Directions for Social Neuroscience." *Neuron* 65, no. 6: 752–67.

Andre, Judith. 2002. *Bioethics as Practice*. Studies in Social Medicine, edited by Allan M. Brandt and Larry R. Churchill. Chapel Hill: University of North Carolina Press.

Bickle, John. 1992. "Revisionary Physicalism." *Biology and Philosophy* 7, no. 4: 411–30.

Bunge, Mario. 1977. "Levels and Reduction." *American Journal of Physiology* 233, no. 3: R75–R82.

Cacioppo, John T., and Gary G. Berntson. 1992. "Social Psychological Contributions to the Decade of the Brain: Doctrine of Multilevel Analysis. *American Psychologist* 47, no. 8: 1019–28.

Callahan, Daniel. 1996. "Can Nature Serve as a Moral Guide?" *Hastings Center Report* 26, no. 6: 21–22.

Casebeer, William D. 2003. "Moral Cognition and Its Neural Constituents." *Nature Neuroscience* 4:841–46.

Changeux, Jean-Pierre. 1981. "Les progrès des sciences du système nerveux concernent-ils les philosophes?" *Bulletin de la société française de philosophie* 75:73–105.

Changeux, Jean-Pierre. 1996. "Le point de vue d'un neurobiologiste sur les fondements de l'éthique." In *Cerveau et psychisme humains: quelle éthique?*, ed. G. Huber, 97–109. Paris: John Libbey Eurotext.

Changeux, Jean-Pierre, and Paul Ricoeur. 2000. *La nature et la règle: ce qui nous fait penser.* Paris: Odile Jacob.

Churchland, Patricia S. 1986. *Neurophilosophy: Toward a Unified Science of the Mind-Brain.* Cambridge: Bradford Book/MIT Press.

Churchland, Patricia S. 1998. "Feeling Reasons." In *On the Contrary*, ed. P. M. Churchland and P. S. Churchland, 231–54. Cambridge: MIT Press.

Churchland, Patricia S. 2002. *Brain-Wise: Studies in Neurophilosophy.* Cambridge: MIT Press.

Cranford, Ronald E. 1989. "The Neurologist as Ethics Consultant and as a Member of the Institutional Ethics Committee." *Neurologic Clinics* 7, no. 4: 697–713.

Descartes, René. 1992. *Méditations métaphysiques, Objections et réponses, suivies de quatre lettres.* Edited by J. M. Beyssade and M. Beyssade. Manchecourt: GF-Flammarion.

Dewey, John. 1922. *Human Nature and Conduct: An Introduction to Social Psychology.* New York: Holt.

Dumit, Joseph. 2004. *Picturing Personhood: Brain Scans and Biomedical Identity.* Princeton: Princeton University Press.

Editorial. 1998. "Does Neuroscience Threaten Human Values?" *Nature Neuroscience* 1, no. 7: 535–36.

Farah, Martha J., and Andrea S. Heberlein. 2007. "Personhood and Neuroscience: Naturalizing or Nihilating?" *American Journal of Bioethics* 7, no. 1: 37–48.

Farah, Martha J., and Paul Wolpe. 2004. "Neuroethics: Toward Broader Discussion." *Hastings Center Report* 34, no. 6: 4–5.

Fins, Joseph J. 2005. "Clinical Pragmatism and the Care of Brain Damaged Patients: Toward a Palliative Neuroethics for Disorders of Consciousness." *Progress in Brain Research* 150:565–82.

Francis, Darlene, Josie Diorio, Dong Liu, and Michael J. Meaney. 1999. "Nongenomic Transmission across Generations of Maternal Behavior and Stress Responses in the Rat." *Science* 286, no. 5442: 1155–58.

Gazzaniga, Michael S. 2005. *The Ethical Brain.* New York: Dana Press.

Glannon, Walter. 2007. "Persons, Metaphysics and Ethics." *American Journal of Bioethics* 7, no. 1: 68–69; discussion W1–4.

Greene, Joshua D. 2006. "Social Neuroscience and the Soul's Last Stand." In *Social Neuroscience: Toward Understanding the Underpinnings of the Social Mind*, edited by A. Todorov, S. Fiske, and D. Prentics. Oxford University Press.

Greene, Joshua D., and Joseph M. Paxton. 2009. "Patterns of Neural Activity Associated with Honest and Dishonest Moral Decisions." *Proceedings of the National Academy of Sciences* 106, no. 30: 12506–11.

Greene, Joshua D., R. Brian Sommerville, Leigh E. Nystrom, John M. Darley, and Jonathan D. Cohen. 2001. "An fMRI Investigation of Emotional Engagement in Moral Judgment." *Science* 293, no. 5537: 2105–8.

Hume, David. 1975 [1739]. *A Treatise of Human Nature.* Oxford: Clarendon Press.

Hyman, Steven E. 2009. "How Adversity Gets under the Skin." *Nature Neuroscience* 12, no. 3: 241–43.

Illes, Judy, Mary A. Moser, Jennifer B. McCormick, Eric Racine, Sandra Blakeslee, Arthur Caplan, Erika C. Hayden, Jay Ingram, Tiffany Lohwater, Peter Mc-Knight, Christie Nicholson, Anthony Phillips, Kevin D. Sauve, Elaine Snell, and Samuel Weiss. 2010. "Neurotalk: Improving the Communication of Neuroscience Research." *Nature Reviews Neuroscience* 11, no. 1: 61–69.

Libet, Benjamin. 1985. "Unconscious Cerebral Initiative and the Role of Conscious Will in Voluntary Action." *Behavioral and Brain Sciences* 8, no. 4: 529–39.

Macintyre, Alasdair. 1998. "What Can Moral Philosophers Learn from the Study of the Brain?" *Philosophy and Phenomenological Research* 58, no. 4: 865–69.

MacLean, Paul D. 1967. "The Brain in Relation to Empathy and Medical Education." *Journal of Nervous and Mental Disease* 144, no. 5: 374–82.

Mahner, Martin, and Mario Bunge. 1997. *Foundations of Biophilosophy.* New York: Springer.

Marcus, Steven J., ed. 2002. *Neuroethics: Mapping the Field, Conference Proceedings.* New York: Dana Foundation.

Mayr, Ernest. 1985. "How Biology Differs from the Physical Sciences" In *Evolution at the Crossroads: The New Biology and the New Philosophy of Science*, ed. D. J. Depew and B. H. Weber, 43–63. Cambridge: MIT Press.

Mayr, Ernest. 1988. "Is Biology an Autonomous Science?" In *Toward a New Philosophy of Biology.* Cambridge: Harvard University Press.

McCabe, David P., and Allan D. Castel. 2008. "Seeing Is Believing: The Effect of Brain Images on Judgments of Scientific Reasoning." *Cognition* 107, no. 1: 343–52.

McGowan, Patrick O., Aya Sasaki, Ana C. D'Alessio, Sergiy Dymov, Benoit Labonte, Moshe Szyf, Gustavo Turecki, and Michael J. Meaney. 2009. "Epigenetic Regulation of the Glucocorticoid Receptor in Human Brain Associates with Childhood Abuse." *Nature Neuroscience* 12, no. 3: 342–48.

Moreno, Jonathan. 1999. "Bioethics Is a Naturalism." In *Pragmatic Bioethics*, ed. G. McGee, 5–17. Nashville: Vanderbilt University Press.

OECD. 2002. *Understanding the Brain: Toward a New Learning Science.* Paris: OECD Publications.

Pontius, Anneliese A. 1973. "Neuro-ethics of 'Walking' in the Newborn." *Perceptual and Motor Skills* 37, no. 1: 235–45.

Racine, Eric. 2007. "Identifying Challenges and Conditions for the Use of Neuroscience in Bioethics." *American Journal of Bioethics—Neuroscience* 7, no. 1: 74–76.

Racine, Eric. 2008a. "Interdisciplinary Approaches for a Pragmatic Neuroethics." *American Journal of Bioethics* 8, no. 1: 52–53.

Racine, Eric. 2008b. "Which Naturalism for Bioethics? A Defense of Moderate (Pragmatic) Naturalism." *Bioethics* 22, no. 2: 92–100.

Racine, Eric. 2010. *Pragmatic Neuroethics: Improving Treatment and Understanding of the Mind-Brain.* Cambridge: MIT Press.

Racine, Eric, Ofek Bar-Ilan, and Judy Illes. 2005. "fMRI in the Public Eye." *Nature Reviews Neuroscience* 6, no. 2: 159–64.

Racine, Eric, Ofek Bar-Ilan, and Judy Illes. 2006. "Brain Imaging: A Decade of Coverage in the Print Media." *Science Communication* 28, no. 1: 122–42.

Racine, Eric, and Judy Illes. 2006. "Neuroethical Responsibilities." *Canadian Journal of Neurological Sciences* 33:269–77.

Roskies, Adina. 2002. "Neuroethics for the New Millennium." *Neuron* 35, no. 1: 21–23.

Safire, William. 2002a. "Neuroethics Belongs in Public Eye." *Dayton Daily News,* May 17, 2002, 13A.

Safire, William. 2002b. "Visions for a New Field of Neuroethics." Paper read at Neuroethics: Mapping the Field, at San Francisco.

Stent, Gunther S. 1990. "The Poverty of Neurophilosophy." *Journal of Medicine and Philosophy* 15, no. 5: 539–57.

Chapter 8

Fast-Moving Objects and Their Consequences

A Response to the Neuroscientific Turn in Practice

Anne Beaulieu

In various stages of the preparation of this book, the four essays in this part were labeled as new voices from the neuroscientific turn. These metaphors invoke sound and motion, and call to mind the experience of the Doppler shift—the changing frequency of a wave, as the source or listener move relative to each other. (This is the sound modulation that any three-year-old will enthusiastically imitate as a fast-moving motorcycle zooms by—neeeeeaaaaooooow.) This dynamic image seems very appropriate, not only because it highlights the fast pace of change in disciplinary configurations in relation to neuroscience, but also because it stresses the need to consider the relation between source and listener in context. In what follows, I therefore offer some reflections on what these essays convey about the neuroscientific turn, and begin to analyze their movement in relation to the neurosciences, humanities, and social sciences.

Certainly, the "neuroscientific turn" is not the first turn to be announced and considered, and it echoes a multitude of other turns. In my field, science and technology studies (STS), a series of turns have been observed by different commentators, among them, the empiricistic, cognitive, cultural, spatial, ethical, and performative turns. But STS is not alone in taking turns: in literary studies the transnational, political, empirical, descriptive, material, cultural, linguistic turns have been hailed, in addition to the semiotic and the postcolonial turns. It would seem that a "turn" is a useful rhetorical tool to signal shifts in approaches and objects of study in a field and to mark out a new territory for a coalition of scholars; the turn evokes shifts away from a tradition and toward new vistas.

One of the distinctive elements that mark the neuroscientific turn is the signature prefix that marks a field or approach—the "neuro" is not just a qualifier added to a noun but a renaming of an existing field or a foundational move, a baptism of a new field, as observed by Littlefield and Johnson in the introduction to this volume. Time and again in this book (Birge, this volume) and elsewhere, the burgeoning of terms using the prefix *neuro* is noted. Indeed, at times, this very burgeoning seems to be reason enough to render these fields suspect (Vrecko 2010; Vidal 2009)—"simply add neuro and stir" being the accusation of superficiality and jumping on the bandwagon. But rhetoric is rarely only superficial, and there is more to the "neuro" as prefix than an attempt to be fashionable.

So what is the movement in the case of the neuroscientific turn, who or what is moving, and, importantly, who is witnessing this move? Is it a case of the neurosciences turning toward the humanities and social sciences? Or the other way around? Who is doing the turning? What is the impetus for such a turn? The four essays in this part provide rich materials to consider what is at stake in a turn, and to explore what a neuroscientific turn might mean for the various modes of inquiry involved.

Let's consider how these essays posit a neuroscientific turn. The first of these argues that while much attention has indeed been paid to the way the neurosciences have influenced the humanities and social sciences, another turn should also be investigated. Wishing to challenge the perceived asymmetry in which the neurosciences are of relevance for the humanities, Birge sets out to consider how neurofiction (novels in which neurological conditions feature prominently) influences the neurosciences. In posing this question, Birge emphasizes the views and context of clinical practitioners. She illustrates how, in the context of discussions about neurofiction, the trope of an "arts/science" distinction to be bridged is put forth. What is gained from this turn includes increased awareness of conditions, and "better" (in the sense of complementary or many-sided and contextualized) descriptions of experience of patients for neuroscientists. Furthermore, fiction engages the mind in different ways, and this may support neuroscientists in the creative aspects of their research.

While not an explicit focus of the chapter, there are some indications as to how the interaction takes place. For example, when the neurosciences turn to fiction, to novels that convey experience of illness or that highlight the moral valence of a practice, this helps doctors to gain a better understanding of human experience. This dynamic is not only found in dedicated medical humanities courses, but also closely related to the humanist tradition of American medical education. Birge also notes that the skills of observation of the clinician may not be so different from those of the novelist, and that both fiction writing and research are creative acts. Here, the (auto)biographical testimonies of a number of prominent clinicians and neuroscientists are the evidence for this crossover. The "turning" is done here by the human agent, who, by becoming cognizant of fiction and of neuroscience, becomes a better clinician or researcher.

In this exploration of the turn, Birge finds a complementarity of neuroscience and neurofiction—one that is also noted by neuroscientists as well as literary critics and scholars whose work she analyzes. As Birge demonstrates, these discussions use explicit metaphors of "gaps" to order kinds of knowledge, or else use implicit assumptions that nevertheless convey that novels are valued in various ways in relation to neuroscientific and neurological knowledge. While this is in itself not so surprising, a close reading of these differences reveals nevertheless that the epistemic commitments of neuroscientists do not seem much altered. Neuroscientists and physicians position these works according to a neuroscientific hierarchy—closeness to correct neuroscientific description is the mark of a better novel. In highlighting these hierarchical elements in the encounter between neurofiction and neurologists, Birge helps to map out the constitution of the uneven terrain where fiction and science meet.

In the second chapter in this part, Hendrix and May propose to develop a "fuller framework for understanding medieval mystic behavior." This kind of knowledge will come from a "neurohistorical analysis," in which historians turn to neuroscience to understand differences in effects of different technologies of the self of mystics. The authors argue that the benefits will be bidirectional and yield fuller insights than either neuroscience or history can provide individually.

The neuroscientific turn, in this case, is achieved by matching descriptions of the practices of mystics, at the level of "technologies of the self," to particular insights from neuroscience. This implies bridging what we know about how these techniques of the self were implemented historically and what is currently known about the neural correlates that result from specific experimental procedures. Such a turn therefore relies on correlating what we know about historical events and what we know about neural phenomena.

The authors note possible issues in making this connection, such as the fact that the two kinds of materials they use arise from very different time periods—historical accounts versus contemporary neuroscience. Yet, these materials can be combined, they argue, because what we know about the brain today is valid for the medieval brain too. This line of reasoning posits that some knowledge about some kinds of objects, such as historical accounts, requires contextualization, while knowledge about brains does not. Furthermore, in arguing that research in "contemplative neuroscience provides a heuristic for a different understanding of ostensibly similar practices by historical figures" and that "neurohistory allows the researcher to consider alternate routes into the subjectivities of historical figures," the authors put forth very specific epistemological and ontological distinctions between these kinds of knowledge. This characterization further reinforces an understanding of historical research as being about representations (words on paper) of cultural contexts, while neuroscience addresses the underlying and unchanging realities of individuals by providing knowledge about the universal brain. This has implications for the "turn," and I will return to this below. But note how this treatment sets up one kind of knowledge as heavily situated and context dependent, while another is universal and timeless. Finally, it is interesting to observe that the neuroscientific materials on which the authors draw to achieve their turn include primary sources, but also popular science monographs and secondary sources—as does the chapter by Gorzelsky.

Gorzelsky's account seeks to develop an approach that would bridge two types of scholarship in studies of literacy: personal changes effected by literate practices and systemic changes brought about by literate practices. She proposes to do this by considering how biocultural environments are changed by interventions in one's own psychophysical processes. This aim actually requires two kinds of bridges, between biological and cultural, and individual and systemic. In order to achieve this, Gorzelsky, like the authors of the other chapters, also embraces the clinical—in this case, through clinical psychology and self-help. She correlates the biocultural situation in Vietnam to that of PTSD victims, in order to draw parallels between therapies developed for patients and literacy practices propagated in Vietnam. Furthermore, she posits a mechanism that would link the biological and the cultural as well as the individual and the social.

Like Hendrix and May, Gorzelsky proposes a neuroscientific reading of specific practices, hoping for a better understanding of how these make a difference for individuals and societies. The chapter therefore seeks to develop an integrative turn, an approach that would enable analysis across different phenomena that are usually treated as separate objects of separate modes of inquiry.

In the chapter by Racine and Zimmerman, the focus is on labeling streams of research in order to make explicit their scopes and ambitions. This exercise supports a discussion of how different kinds of knowledge intersect. It also makes explicit what the authors see as frameworks that may or may not make it possible for neuroscientific findings to matter in ethical discussions.

Clearly, these chapters therefore do not all point to a single kind of neuroscientific turn but rather propose shifts that are different in their goals and dynamics, putting forth the desirability of mutual influence of different spheres, a complementary interaction between neuroscience and humanities, or a far-reaching integration to seamlessly move from biology to culture and from individual to society.

There are important common elements, however, in the turns proposed. As noted above, the authors use a range of materials from the neurosciences. Editorials and opinion pieces, secondary sources, popular science sources, reviews and meta-analyses as well as "original research publications" are all called upon to stand in for neuroscientific knowledge. In some instances, this variety of materials is used to trace the dynamics of neuroscientific knowledge in both science and culture and in the wealth of texts that arise from contemporary interest in the brain. Other chapters however rely on this material as providing a hard ground of universal knowledge about the brain. This raises a number of crucial questions about the kind of disciplinarity that is being created by this scholarship. Does the neuroscientific turn necessarily mean that the ontology and epistemology of neuroscience have to prevail? And what kind and what degree of expertise are required to delve into neuroscience, which, like any other academic specialty, has its own specialized language, empirical approach, and analytical style? I would argue that in order to understand neuroscientific knowledge and its potential significance for other kinds of research, it is essential to be aware of its context of production and interpretation. By extending to neuroscientific knowledge a special status in a priori and blanket fashion, the risk is to reinforce the hierarchy and asymmetry in the treatment of different types of scholarship.

For a scholar trained in science and technology studies, the diversity of sources used in these chapters is problematic. There is a tendency to put popular science books on par with "original research articles." It is of course very valuable to consider the role and import of popular scientific literature on the brain, and how this discourse is itself part of the neuroscientific turn. But different kinds of texts have different roles in research as in culture. To treat all these texts as equally authoritative, and as being authoritative in the same way, is to be remiss in the deployment of critical approaches to knowledge creation. This move of putting on par different kinds of texts levels a whole series of distinctions that are crucial to the ways knowledge is produced and validated in the neurosciences. The same goes for the disciplinary distinctions and experimental traditions that get mixed in the course of the various chapters.

Am I then arguing that humanities scholars should become neuroscientists? Am I trying to be the science police, demanding that there be one correct understanding of neuroscientific work? Am I pushing for a colonization of the humanities by neuroscience? Far from it! On the contrary, these essays signal a fascination and curiosity for the neurosciences that are important to develop. But my concern is that in turning to neuroscience, in seeking to enrich their understanding, scholars need to be careful not to lose sight of what it means to do humanities scholarship. To return to my point about the diversity of sources used, it seems to me that thinking very hard about the contextual understanding of knowledge and the critical consideration of sources and texts are core values in the humanities. More attention to the variations in the kinds of neurosciences that are out there, and how best to use different sources, should be important practices in the course of a neuroscientific turn. In turn, the humanities are equally rich, with substantial scholarship on many issues that are becoming of interest to the neurosciences. Clearly, one of the dangers for the neurosciences is that if too little attention is paid to this scholarship, the treatment of issues will remain superficial, or a lot of energy will be expended on reinventing the wheel. Racine and Zimmerman, in their careful mapping of the various strands of neuroethics, show the differences that arise when ethics is discovered anew by neuroscientists and when it is developed in conversation with both philosophy and neuroscience.

Furthermore, paying attention to the context of production of this knowledge does not automatically mean taking over the epistemics of neuroscience. On the contrary, humanities scholars can approach the scientific literature in their own way, using their own criteria for valuing particular research and writings—and these need not be the same as those of neuroscientists. But such a move would equally carry the obligation to make those criteria explicit and to be able to respond to queries about selection of particular insights from the neurosciences. The openings of the chapters by Birge and Gorzelsky, where the authors sketch the desirable features of an arena where insights from different disciplines can meet, are excellent steps in this direction. By making clear why and how certain insights have been selected, the work will be more recognizable to other humanities scholars who may be more distant from the turn. It will also prevent humanities scholars from being accused of "picking and choosing" across the huge neuroscience literature—a position of weakness in the eyes of neuroscientists that is best avoided if the humanities are indeed to be a valued partner in the exploration of human experience.

Another element common to these four chapters is that they all draw on material from clinical science. In science and technology studies, such differences between clinical, research-based, and lab-oriented practices tend to be carefully mapped out and distinguished, since they represent distinctive traditions and commitments. What is the role of the strong presence of "the clinical" in these accounts? Not having had the privilege of conversation with the authors, this must be taken as a question raised rather than a conclusion. So why is the clinical a useful way into neuroscience? Perhaps the figure of the patient is part of the answer. In clinical discourse, the subject is given a place, and generally, so are narrative and intersubjectivity. Although there are notable exceptions (Roepstorff and Jack 2004; Roepstorff 2001), more experimental writings in neuroscience tend to systematically censor these elements. Given that analyses of forms of subjectivity (including agency), narrative, and intersubjectivity are core elements of humanities scholarship, they may function as beacons of recognizable discourse in the clinical literature, helping to bridge some of the gaps between the epistemic cultures (Knorr-Cetina 1999) of humanities scholars and of neuroscience. Here too, humanities scholars might do well to hold on to the concept of the subject (and all the problematization that accompanies it) as a core competence and sensibility of their field. Indeed, by making explicit why such a concept is needed to understand human experience, and by querying the causes and consequences of its absence from neuroscience, the humanities could develop their own relation to the neurosciences and put forth a precious contribution arising from their own core epistemics.

Further reflecting on the kind of scholarship presented in this part, it seems that texts and published literature are indeed at the center of the turn as presented in these chapters. Clearly, humanities scholars are reading new kinds of materials, whether neuroscience books or medical journals. But other bases for interaction between the neurosciences and the humanities and social sciences are also possible—for example, explicit collaboration, or coauthorship (which is the case in the chapter by Hendrix and May), or formal training in the other field. Such interactions, insofar as they would take on more informal aspects than interacting with published literature, would raise even more questions about the new disciplinary formations that may be developing.

If the neurosciences are a fast-moving object whizzing by, there is no obligation for the humanities to be a stationary observer. The possibility of reflexive and explicit frameworks for selection of and engagement with neuroscientific work evoked above can contribute to a more active and accountable combination of humanities and neurosciences. The awareness that knowledge is neither innocent nor immanent should therefore spur the development of strategies to bring together different bodies of work, whether from humanities or neuroscience, that critically interrogate each other rather than seek seamless integration.

NOTE

The author would like to thank the authors in this part for sharing their work in progress and the editors for their feedback on an earlier version of this response.

WORKS CITED

Knorr-Cetina, Karin. 1999. *Epistemic Cultures: How the Sciences Make Knowledge.* Cambridge: Harvard University Press.
Roepstorff, A. 2001. "Brains in Scanners: An Umwelt of Cognitive Neuroscience." *Semiotica* 134:747–65.
Roepstorff, A., and A. I. Jack. 2004. "Trust or Interaction?" *Journal of Consciousness Studies* 11, no. 7: v–xxii.
Vidal, Fernando. 2009. "Brainhood, Anthropological Figure of Modernity." *History of the Human Sciences* 22, no. 1: 5–36. doi:10.1177/0952695108099133.
Vrecko, Scott. 2010. "Neuroscience, Power and Culture: An Introduction." *History of the Human Sciences* 23, no. 1: 1–10.

Part 3

Critical Responses to the Neuroscientific Turn

Whereas part 1 introduced relevant histories and contexts for neuroscientific knowledges and part 2 explored the productive potential for emergent neurodisciplines, part 3 addresses several challenges and key issues for the neuroscientific turn: the influence of emergent neurodisciplines on public policy decisions (Rosenquist and Rothschild), the hype surrounding imaging technologies (Fitzpatrick), the position of neuroethics as interlocutor for the turn (Whitehouse), and the place of neuroscience in the humanities classroom (Michelson). The essays in this part temper optimism with skepticism by questioning some of the basic assumptions that have made transdisciplinary scholarship seem possible and promising. Rosenquist and Rothschild's essay asks social science fields—specifically neuroeconomics—to balance the descriptive benefits of neuroscience with its potential to influence public policy decisions. By delineating the complexities of imaging technologies, Fitzpatrick reminds us that specialized knowledge constructs a border (however permeable) between disciplines. In his exploration of Alzheimer's research, Whitehouse questions the ascendance of neuroethics as *the* key interlocutor for the neuroscientific turn, proposing instead a more nuanced and "deeply contextualized" perspective that tempers the magic of science with the "hard social work" of the coming decades. Finally, Michelson employs a plethora of examples, from William James to William Gibson, that illustrate the intricacy and value of humanities scholarship to any endeavor bent on explaining concepts such as "creativity" and "consciousness." In their sanguine suspicion, James Niels Rosenquist and Casey Rothschild, Susan Fitzpatrick, Peter Whitehouse, and Bruce Michelson remind us that the neuroscientific turn is neither inconsequential nor inevitable.

Chapter 9

Neuroeconomics

A Cautionary Primer

James Niels Rosenquist and Casey Rothschild

Neuroeconomics is a term that has received significant media exposure over the past decade (Bonanno et al. 2008). It has been hailed as a discipline that can answer a number of fundamental questions in the area of economics, while it offers up a new set of findings that can augment and potentially alter standard economic practice (Camerer, Loewenstein, and Prelec 2005). It is, along with its close cousin behavioral economics, one of the fastest growing subdisciplines in the field and even has its own listing in the *Journal of Economic Literature* classification system.[1]

While neuroeconomics (NE) has received its fair share of positive press,[2] a number of economists have challenged the relevance and validity of this work (Gul and Pesendorfer 2009). While detractors identify NE questions as potentially interesting, they often conclude it is not "economics" in the sense that it does not fit within the classical framework of economic research, or, if it theoretically could, it has yet to live up to its potential. Furthermore, some argue that the public policy suggestions proffered under the auspices of NE research—for example, optimizing "neural utility"—may be both unjustified by the NE research and fundamentally misguided (Bernheim 2009).

The goal of this chapter is to present an overview and some cautionary words about neuroeconomics. There are already a number of reviews of NE within the economics literature, from the enthusiastic (Camerer, Loewenstein, and Prelec 2004) to the pessimistic (Rubinstein 2008) to the ambivalent (Bernheim 2009).[3] Our goal is not to add to that literature, which is written for trained economists; we will instead consider NE from a broader perspective. This is an important exercise for at least two reasons. First, since NE is based upon (at least) two distinct fields of study with disparate semantic traditions, there is a danger of noneconomists misinterpreting the meaning of terms such as *utility*, *positive*, and *normative*. These words have distinct meanings to economists that may not be entirely clear to those with limited formal economic training.

Second, and perhaps more important, NE research has the potential to broadly influence public policy; it is therefore crucial that both its strengths and potential shortcomings are understood beyond the narrow confines of academic economics. To this end, we explicitly compare and contrast the role of economics in guiding public policy with the role of basic medical science in guiding medical interventions. Both fields rely upon "basic" research as to how the system under study (e.g., a particular market or the human body) will function with and without an intervention. When coupled with a *value judgment* about what "proper" or "impaired/diseased" functioning is, this basic research can guide prescriptive economic policy or health prescriptions.

In the following sections, we will present neuroeconomics as a discipline in the context of its historical roots in traditional economics and behavioral economics (BE), consider its contributions to economics, and speculate on what its future might hold. We will argue that NE, as an economic subdiscipline, may provide useful *descriptive* insights into how economic decisions are made, and that it can help to clarify the types of value judgments that need to be made to inform public policy. We will argue that it cannot, however, serve as a substitute for these value judgments. On the contrary, behavioral economics and particularly neuroeconomics imply an essential and central role for making explicit value judgments in deriving public policy prescriptions.

Economics Defined

The field of modern economics has generally been defined as the branch of social science that deals with the production, distribution, and consumption of goods and services.[4] From its origin, economics has been a highly mathematical discipline, relative to the other social sciences, and has sought to generate testable hypotheses based upon a relatively small number of assumptions about the behavior of individual people. Two core assumptions are (1) the identification an individual with his or her desires, or *preferences*, and (2) the notion that these preferences are *revealed* through the choices that individual makes. Coupled with a few auxiliary assumptions, these two core ideas allow economists to model individuals as if they are attempting to maximize a well-defined *utility function*, which weighs different decision options, such as to spend or save. Thus the "self," as defined by economists, is simply the set of preferences—represented mathematically as that person's utility function—that guide their decisions over time, and the most informative guide to these preferences is taken to be precisely the decisions those individuals make.

These assumptions, which until recently were canonical within economics, reflected the core idea that an individual's choices reflect his or her interests, and that individuals are the best arbiters of their own desires. This "individual sovereignty" notion was a value judgment that traditionally was taken for granted in economics. It provides scope for and facilitates the formulation of *prescriptive* (referred to as normative) policy in two ways: (1) identifying ameliorative policies for "market failures," in which social or market institutions interfere with allowing individuals to achieve their goals and (2) formalizing distributional concerns when the desires of different individuals are in conflict, for example, through models such as those for "optimal taxation" (Mirrlees 1971).

The fields that compose the medical sciences provide a useful analogy to the first type of prescriptive economic policy. In the medical sciences, basic scientists explore how the human body operates in healthy as well as altered or "diseased" states. Based upon these findings, biological interventions can be developed that use observed pathways to alleviate diseased states and ensure better bodily function overall. For example, a physician might diagnose an individual as being iron deficient and recommend supplements or dietary changes to cure the individual.

Similarly, traditional "basic" economics researchers seek to identify conditions that will lead to markets that function well or poorly at facilitating individuals' ability to achieve their goals. Positive economic models can then be used to determine how markets may be manipulated to remediate poor functioning of particular markets. For example, an economist might observe that an individual who drinks alcohol and drives a motor vehicle imposes costs not just on himself but on other drivers around him, so that a "free market" in drinking and driving will lead to bad outcomes (a market failure)—outcomes with an excessive number of accidents and deaths. Intervening by banning drinking and driving could mitigate this market failure, leading to an outcome in which every individual is better off.

A major difference between medical interventions and traditional economic ones is that the former identifies diseases *of* individuals and makes recommendations about possible remedies those individuals might choose to pursue, while the latter assumes identified pathologies that arise from the interactions *among* individuals and directly intervenes to correct those pathologies *for* "them" as a collective group. Because the treatments are *of* individuals in the medical context, recommended treatments could, in principle, be refused by the patient (e.g., a Jehovah's Witness or a Christian Scientist). In the "market failure" context, individuals would typically neither want nor need to have the option to refuse "treatment": by construction, the intervention is designed to make them better off, as they themselves perceive "better." Put another way, the only "true" disease to traditional economists is when people cannot choose what they want, and therefore they perceive no need to "ask permission" to implement the cure. As we discuss below, the rise of behavioral economics has expanded the conception of what can be diseased to include the possibility that individual decision making can itself be "diseased."

The Rise of Economics and the Challenge of Behavioral Economics

In order to understand the importance of NE as it relates to both economics and society, it needs to be presented in the context of recent events that have expanded the influence of economics while challenging the core assumptions of traditional economic models. The first key change related to the expanding scope of questions being studied (and policies being framed) in an economic context. Models originally applied to the consumption and savings decisions of individuals, or the hiring, production, and investment decisions of firms, turned out to be flexible enough to apply to a much broader set of problems, including individual-level decisions about "investments" in education, marriage, and the use of addictive substances. Indeed, any problem that can be modeled as that of an individual with well-defined goals and beliefs who is making choices in the face of constraints can be fruitfully analyzed with the same basic model (which Gintis [2008] succinctly refers to as the "beliefs, preferences, and constraints" model). The simplicity and adaptability of these economic models fostered a period of growth into areas that were previously thought of as being part of sociology or other disciplines (Levitt and Dubner 2005).

While the field of economics was growing in terms of its scope and influence, another set of research findings highlighted the weakness of some of traditional economic assumptions related to preferences. Traditional models assumed that an individual's preferences were unchanging over time, and that an individual's choices would be consistent regardless of the way in which the choice was presented. While expected utility and related traditional models such as rational choice theory[5] were useful and generally predictive of economic behavior, they were less able to explain a number of economic choices made by people. For example, individuals appear to have preferences that vary from one month (or week, or day) to the next and often exhibit "time inconsistencies," whereby their perceived trade-offs between, say, January 1 and January 2 depends strongly on how far in the future January is. Another finding that goes against traditional economic assumptions is that people consistently show different preferences depending upon the *way* a choice is described to them (referred to as the *framing effect*). These anomalous outcomes, linked with previous work in the field of psychology (Tversky and Kahneman 2000), contradicted many of the long-standing assumptions about how individuals make economic choices. Alternative models of economic behavior, most famously prospect theory (Kahneman and Tversky 1979), began to receive significant attention when they were found to be better predictors of behavior than traditional models in a number of real-world settings such as employee enrollment in 401(k) retirement plans (Choi, Laibson, and Madrian 2004). Mainstream acceptance of these models, which formed the foundation of a new field known as behavioral economics (BE) was evidenced by the awarding of the Nobel Prize in economics in 2002 to Daniel Kahneman and his colleagues for their work on prospect theory. Thus, while economics as a field was expanding its influence in the social sciences and public policy, its core tenets were being challenged from within.

Criticisms of Behavioral Economics and the Rise of Neuroeconomics

Though the behavioral approach has gained traction in economics, there are a number of criticisms of its approach (Rubinstein 2006). One common critique is that, although BE models may in some cases be more accurate in their predictions than traditional models, they may not applicable in all cases—that is, they are not easily generalizable across distinct contexts. Implicit in these criticisms is that BE models are based upon anomalous psychological findings that cannot be fully interpreted in a concrete and cohesive way. Traditional models, on the other hand, despite their faults, are at least based upon a set of consistent, unifying, assumptions.

Another critique of BE concerns its relation to normative policy prescriptions. While behavioral economics has led to significant improvements in a descriptive sense, it has also implicitly challenged traditional public policy recommendations by undermining the simplicity of the core value judgment of traditional economics. For example, traditional models assume that there is a single economic "self" in the sense that an individual's preferences do not change over time. If an individual does not have consistent preferences over time (a finding or assumption that is implicit in much of the BE literature) or makes different decisions depending upon the way that the choice is presented, it is no longer clear which set of preferences (or self), if any, policymakers should seek to maximize through public policy.

Returning to the analogy with medicine, BE models open up the possibility that *individuals* can have systematic biases in the way they make choices: what was once thought of as a healthy (freedom of choice) state can now be seen as diseased. This poses some new and deep challenges. In most medical interventions, individuals are diseased, but they ultimately have the right to accept or reject treatment for that disease. Traditional economic policy interventions, on the other hand, are based on ameliorating market failures to help markets better allow individuals to achieve their own, self-defined goals; for that reason, individuals at least tacitly "give consent" to these interventions. But in BE models, the "disease" calling for intervention involves a (supposed) pathology *in the way individuals make choices*, for example, when things that "should not" matter, like framing, nevertheless appear to. In that situation, it may not be possible for individuals to "consent" to the sorts of policy interventions that would cure this pathology.

As BE models have been met with resistance and criticism from both economists and psychologists, neuroeconomics arose in part to provide a more concrete foundation for BE models as a defense against this criticism. In particular, it arose as economists turned their focus on identifying the more fundamental processes guiding decision making within the brain, with the goal of linking those processes with the "higher level" findings of BE (and economic behavior more broadly) (Camerer, Loewenstein, and Prelec 2007). This approach sought to link what is often referred to as the "basic" field of neuroscience with the social science of economics and hence became known as neuroeconomics.

An understanding of neuroeconomics is thus quite relevant to the ongoing academic and policy debates between traditional and behavioral economics since it can (and does) present evidence from basic neuroscience research as a "hard" scientific foundation for the insights from behavioral economics. NE research thus has the potential to provide a scientific foundation (or at least gloss) to the *descriptive* and *predictive* elements of behavioral economics.

Neuroeconomics Defined

The New Palgrave Dictionary of Economics states, "Neuroeconomics aims at improving the science of major economic phenomena. . . . A revised model of choice is expected, based on the behaviour of the neuronal structures of the brain" (Dickhaut and Rustichini 2008). While this description is broad (and arguably speculative), it is a reasonable description of what NE researchers seek to do. NE researchers focus primarily on the antecedents of economic choices through a variety of experimental methods common in neuroscience and cognitive psychology, including functional magnetic resonance imaging (fMRI), that allow researchers to measure localized brain activity in real time (McClure et al. 2007), single-cell recordings in animals that can be used to study basic learning behavior at the cellular level (Glimcher 2009), and noninvasive measures of brain activity including galvanic skin response and electroencephalogram (EEG) measurements that can measure arousal and brain activity. In addition to animal models, experiments can assess how various "treatments" (Zak and Fakhar 2006) may impact economic behaviors and outcomes.[6] The goal of these studies is to test and inform theoretical models of economic behavior, which can then be further refined with empirical analyses.

How Neuroeconomics Can Inform Economics

As is often the case with new disciplines, speculation about potential contributions is usually more wide-ranging than a discussion of actual findings. This is certainly true of neuroeconomics (Camerer 2007). There is a wide array of potential uses of NE research that fit broadly within positive and normative economics.[7] As discussed above, positive economics is concerned with the predictive ability of any given model of economic behavior, while normative models seek to prescribe welfare-improving public policies. NE will make a contribution to positive economics insofar and only insofar as it improves the descriptive (and predictive) power of economic models of behavior, that is, of how individuals *actually* make choices. As we discuss further below, the role NE has to play in normative economics is far muddier.

Positive Neuroeconomics

For NE to be useful in positive economics, it simply needs to generate models of behavior that are an improvement over those currently in existence.

This could conceivably relate to predicting variation in an agent's behavior over time or variation between individuals. For the purposes of this chapter, we will focus on NE's potential to describe within-person variation only.[8] A clear example of how NE could prove useful in this context relates to the idea of preferences that are inconsistent across different moments in time. One approach to modeling this in the BE literature has been to impute different economic "selves," each with distinct preferences, who act at different times within a given individual. While it cuts quite strongly against canonical economic modeling, the idea that individuals are "really" composed of various *competing* impulses is hardly a new one; think of the representation of the angel and devil on one's shoulders seeking to influence behavior in different (and often opposing) directions, Whitman's famous observation that he "contains multitudes," or Freud's writings on unconscious conflict.

NE research can potentially shed light on what this notion of multiple selves "really" means in practice. For example, McClure et al. (2004) present fMRI evidence that distinct parts of the brain are activated depending on whether a reward is immediate or in the future, which suggests that different parts of the brain are responsible for short- and long-term rewards. This experimental work (and other work like it) served as the foundation for Bernheim and Rangel's (2004) model of addiction. In this model, cues (such as beer advertisements) that elicit a strong emotional response can transform an individual from a "cold" to a "hot" self that wants to use an addictive substance—that is, can shift the locus of the part of the brain actively making decisions from the part identified by McClure as being associated with long-term rewards (the lateral prefrontal and posterior parietal cortex) to the part identified as being associated with the short-term rewards (the midbrain dopamine system). This model better predicts behavior among addicts than "rational models" (with a single self) in a number of instances such as inpatient rehab (rational addiction models suggest that self-imposed hospitalization would be unnecessary since people would simply choose to stop using). It is reasonable to assume that other models based upon NE approaches could also prove useful to the field of positive economics suggesting that, while NE is currently limited in its usefulness, it does provide some clear examples of how it might inform descriptive economic models.

Normative Neuroeconomics

While positive NE models can be judged by relatively straightforward criteria, normative NE models (much like normative BE models) cannot be evaluated so easily. This is because, as mentioned previously, normative economic models require value judgments. When individuals have well-defined and consistent preferences, they define a natural set of value judgments: helping the individual achieve what he or she wants. This is the basis for normative models in traditional economics. In light of neuroeconomic and behavioral evidence that individuals do *not* have well-defined and consistent preferences, there is less clarity about—and consequently broader scope for—value judgments.

That is, because behavioral economic research has suggested that the notion of "what individuals want" is less clear-cut than traditional economics assumed, it has undermined the commonly accepted value judgment underlying the policy prescriptions of traditional economics. At the same time, it has broadened the potential scope of economic policy interventions. For example, there is no scope in traditional economics for "correcting" individual decision making—only *collective* decision making—but it is possible to contemplate interventions designed to "cure" individual decision-making "failures" within the behavioral and neuroeconomic paradigms.

We view this broader conception in economics of the scope for public policy as necessary, but we also view it as potentially quite dangerous. Economists are traditionally accustomed to tacitly agreeing on the value judgments underlying their policy prescriptions. They are therefore not accustomed to making the value judgments underlying their policy prescriptions explicit. In traditional economics, to say that there is a market *failure* is to make a precise statement resting on a concrete but usually only tacit value judgment: what is "good" is what *individuals* want, and markets that do not successfully promote this "good" are pathological and should be "cured."

Behavioral economics frequently uses similar pathologizing language to refer to individual choices. Individuals might be identified as "present *biased*," for example, suggesting that individuals have pathological tendencies to choose to consume too much and save too little. The word *bias* (and hence the policy prescriptions suggested to correct it) clearly involves a value judgment, but it may not always be clear what that value judgment is. It could, for example, be something along the lines of "We should value choices made under long reflection more than choices made impulsively in the heat of the moment." Alternatively, it could be something like "Patience is a virtue." Either could form the basis for a policy recommendation of subsidized or even coerced savings to "correct" the bias, but they nevertheless have quite different flavors: the former suggests that such a recommendation will help individuals do what *they themselves* "really" want (if only they reflected for longer); the latter suggests that it will help individuals do what the *prescriber* views as desirable. Our concern is that unless value judgments are made clear and explicit, the latter types of value judgment can masquerade as the former. That is, under the guise of "curing" some pathological "bias," economists could prescribe policy interventions that essentially amount to coercing individuals into doing what *the economists* think is best.

For example, there is obviously a tacit value judgment being made in the angel-versus-devil conception of internal conflict. Bernheim and Rangel (2004) apply a similar value judgment in their model of cue-triggered use of addictive substances, discussed above: they explicitly make the value judgment that the desires of the "cold self" (i.e., the "deliberative" part of the brain activated by long-term rewards) are what really matters. (Other value judgments are possible, of course, such as one that weighs both the hot and the cold self equally, but one can only draw *normative* conclusions from this model by imposing some concrete value judgment about what is good or bad.)

It is useful to clarify the precise role of NE evidence in this example: it formalizes the notion of "impulsive" urges by identifying a consistent pattern of brain activity that is associated with that informal notion. This allows greater precision in describing those times when the modeler or policymaker judges particular paternalistic interventions to be ethically justified: one can say, for example, that one regards paternalistic subventions of individual free choice as justifiable whenever the brain is displaying a certain pattern of activity, here dubbed the "hot state." What the NE evidence does *not* do here is fundamentally alter the basic value judgment implicit in those paternalistic interventions.

In other words, the NE evidence allows us to convert an imprecise ethical judgment, such as "It is okay to intervene in case an individual is 'impulsive' or 'out of control,'" to a precise one like "it is okay to intervene in a 'hot state' as defined by the following types of brain activity." It does not—and, indeed, cannot—ever lead to is-ought statements like "It is okay to intervene *because* he is in a hot state."

The risk posed by normative neuroeconomics is that it can be used to obscure—indeed, to obfuscate—the implicit value judgments involved in reaching public policy conclusions. For example, a policymaker might hold the view that addictive substances (or fatty foods, or a lack of exercise, etc.) are *bad things* in and of themselves. This policy maker would therefore be inclined to label the "hot state" identified by neuroscience as being associated with use of these substances as bad *because* of one's underlying value judgment about the behavior it implies. In this case, the neuroscientific evidence is orthogonal to the *real* value judgment, and, if used in support of a policy intervention, would be doing little more than obscuring the role of that judgment.

This risk is exacerbated when NE findings are used to buttress preexisting models from the BE literature. For example, data typically interpreted as evidence of "optimism bias" and "present bias" in the context of adolescent addiction can in some circumstances be equally well be interpreted as evidence of "affinity preferences"—whereby adolescents are unbiased but are uncertain about who they will grow up to be, and care more about some future selves than others (as suggested by Rosenquist and Rothschild [2011] as well as Bartels and Rips [2010]). More generally, our view is that we simply do not yet have firm enough understanding of "high level" models from the behavioral literature to feel confident treating "low level" NE evidence as a "smoking gun" in favor of one particular model and its associated normative valence.

Economics is known for its rigorous analysis of data and spirited, intense debates over theoretical and empirical matters.[9] We are concerned that it has to this point remained largely silent when presented with fMRI images of activity in the cortex that are cited as evidence of BE models of time discounting (McClure et al. 2004). Whether this relative quiet is due to unfamiliarity with the particular scientific methods being used, some version of the vividness effect (from psychology) where bright colors have been shown to be more convincing to reviewers than those presented in black and white, or some other reason, is unclear. What is clear is that NE research, which in a number of cases seeks to link single-cell recording data or fMRI results with economic behavior, is still some distance from being usefully cited as evidence strongly supporting particular restrictive sets of assumptions. For example, there is still significant disagreement among behavioral researchers about whether the psychological processes cited by BE researchers have *ever* been satisfactorily linked with neuroscientific processes (Williams and Taylor 2006). A specific, related set of critiques of fMRI technology have also appeared within the past few years.[10]

A final concern we have about the normative applications of NE research is more speculative. Some economists view NE as having the potential to operationalize Edgworth's "hedonimeter"—a device that directly measures the pleasure or pain (utility) of an individual (Colander 2007) and can therefore be used to maximize that individual's "true" utility. Dopamine activity in the nucleus accumbens (sometimes referred to as the "pleasure center" of the brain) has been suggested as one possible direct measure of pleasure (Caplin 2010).[11] Monitoring dopamine activity and formulating policy interventions designed to maximize it strike us as alarmingly dystopian (think of Soma in Huxley's *Brave New World*).[12] To be sure, no one that we know of is suggesting that we formulate policy in this way. Nevertheless, neuroeconomics opens up scope for movements in this direction, and these movements are not without their dangers.

In sum, while NE may be useful in improving positive models of economic behavior, its direct role in the normative is less clear, more limited, and potentially dangerous. NE does not and cannot provide evidence for or against particular value judgments made in designing public policy. What it *may* be useful for in the normative is by helping to clarify the value judgments implicit in reaching particular conclusions. Knowing that there is a neurobiological hot state that is cued by certain triggers may, for example, make us feel more comfortable staying the hand of an alcoholic friend as he reaches for a beer the morning after deciding to quit. In that case we can express our own values more parsimoniously: we value the cold-state "self" more than the hot-state "self"—and know, from the scientific evidence, that these mental constructs are potentially sensible ones.

Assessing Neuroeconomics and the Road Ahead

While there is general disagreement about neuroeconomics' potential for making important contributions to economic understanding (and for productively informing public policy debates), there is a clear consensus among economists that the field's major contributions—if any—are as yet unrealized: the current scientific methods used in this research is nascent and, as such, has nontrivial limitations. While this "wait and see" attitude is shared by the vast majority of economists, it is quite easy at this time to draw (what we view as) overly hasty conclusions about normative policies if one is not vigilant.

In the book *Nudge* (2007), Cass Sunstein and Richard Thaler use BE and NE research to suggest a set of "nudge" policies that promote particular policy goals without limiting freedom of choice. Retirement savings is a central example. Evidence strongly suggests that new hires who are, by default, enrolled in a 401(k) plan are much more likely to participate in that plan relative to those who are, by default, not enrolled, even when enrolling or unenrolling from the plan involves trivial effort (Choi, Laibson, and Madrian 2004). Since it leads to higher levels of retirement savings, Sunstein and Thaler are led to promote a policy of implementing an "opt-in" default. On its face, it is hard to argue with this sort of "soft" paternalism—after all, it affects, but doesn't restrict, individual choice. (Of course it is not entirely clear how one reaches a conclusion about which particular choice should be encouraged by the default policy—see Colander and Chong 2010.)

It is worth being clear about what the slippery slope implicit in promoting this sort of soft paternalism entails, however. Promoting "softly" paternalistic policies 401(k) opt-out defaults requires making two (potentially competing) value judgments: an ex ante value judgment that people save too little and a general precept that free choice is good. But once this is granted, the particular choice of a purely soft-paternalistic intervention is not uniquely compelling: it involves valuing (even the slightest bit of) additional free choice by more than one values *any* level of increased saving. Once one has already granted that both things have value, it is, perhaps, more natural to imagine a richer trade-off between the two. For this reason, it is not surprising that advocates of soft paternalism typically see it as a special case of a more broadly paternalistic agenda.

For example, George Lowenstein and Peter Ubel (2010) argue in a recent *New York Times* editorial that to truly achieve policy goals such as the reduction of obesity, soft-paternalism policies need to be augmented by traditional, "hard" paternalism. They cite examples that would limit the choice of individuals in the short term (such as by raising taxes on junk food) that would presumably improve the same individual's long-term health. Debates over the appropriate scope of "hard" governmental paternalism are not new, and many people firmly believe that government has an important role as a benevolent, paternalistic agent. What *is* relatively new is the use of neuroscientific evidence to argue there is a *scientific* basis for paternalistic interventions. For example, Thaler and Sunstein note in a magazine interview that "the human brain is amazing, but it evolved for specific purposes, such as avoiding predators and finding food. Those purposes do not include choosing good credit card plans, reducing harmful pollution, avoiding fatty foods, and planning for a decade or so from now" (2008). In another example, Ubel states in his book *Free Market Madness* (2009), "Markets need to be restrained. . . . When people operate, their unconscious behaviors too often cause them to act against their own interests" (xi). In both cases, the authors cite neuroscientific evidence to support their argument that there are separate economic "selves," and both tacitly assign greater weight to some of these selves than others without making these underlying value judgments explicitly. It is precisely this sort of appeal to neuroscientific evidence as a *substitute* for more direct value judgments that makes us nervous.

There is certainly evidence in the biology literature suggesting an evolutionary antecedent to addiction, overeating, and the desire to spend versus save (Williams and Taylor 2006). But these findings are *positive, descriptive,* findings and do not, in and of themselves, carry any normative weight. Does the fact that our "primordial" selves want to eat a diet rich in fats and sugars make it *bad* for us to choose to do so? Such arguments suggest a *pathologizing* of normal neurobiological functioning and imply the need for the *prescriptive treatment* of public policy, which can ensure the "healthy, cold" self within each of us can be assisted at the expense of the "unhealthy, hot" primordial one. Whether or not one agrees with this prescription, it should be clear that it does not follow in any value judgment–free way from the neuroscientific evidence.

At best, what the NE evidence does is help reframe the nature of our value judgments. For example, we might decide that we value those decisions that *would* be evolutionarily adaptive in *this* world—choices low in fats, sweets, and salty foods—over those that are maladaptive here. At worst, it serves merely to distract from and obscure the underlying value judgments—that is, the author's judgments about what is "really" in a person's best interests.

Conclusion

When Dr. Rosenquist is not working as an economist, he serves as an outpatient psychiatrist, treating people with a variety of serious mental illnesses, including addiction, depression, and attention-deficit/hyperactivity disorder. He listens to his patients' histories, and, through observation of their behavior, attempts to diagnose the cause of their symptoms. While neuroscience has informed him about how their symptoms are caused by their brains' function, and pharmacology has informed about how a certain medication may moderate their symptoms, neither of those fields can provide him with unequivocal proof that the treatment he can prescribe (sometimes against their will) is *better* for them.

As neuropsychiatric evidence has helped people such as Dr. Rosenquist to understand how his patients' brains function in health and in disease, NE has the potential to inform economists by showing, through microfoundational evidence, *how* economic decisions are made. Much as psychopharmacology has developed ways in which to modify brain functioning through medications, NE may very well be able to indicate how behavior *could* be changed by targeting policies in the context of NE models of behavior. What neither psychiatric research nor NE (nor any science for that matter) can ever answer is the question of what is unequivocally *best* for someone; this requires placing our own values into the equation, no matter what the fMRI scan may show.

Neuroeconomics therefore both holds great promise for economics and at the same time raises novel challenges. Economists have traditionally equated what is "best" for an individual with what that individual *chooses* himself. Neuroeconomics will improve our understanding of economic phenomena only insofar as it refines our understanding of how individuals *actually* make choices—and the ways in which this departs from the traditional paradigm. For it to be useful in informing public policy, then, will require economists to think carefully and explicitly about a set of value judgments that they are unaccustomed to making: value judgments about individual choices. The major risk, in our view, is that these necessary value judgments will be suppressed, hidden, and deliberately or unintentionally obscured, and that this will lead to reckless public policy prescriptions.

NOTES

1. See the *Journal of Economic Literature* (JEL) Classification System available at http://www.aeaweb.org/journal/jel_class_system.php#A.

2. See, for example, John Cassidy's essay "Mind Games: What Neuroeconomics Tells Us about Money and the Brain" in a 2006 issue of the *New Yorker* available at http://www.newyorker.com/archive/2006/09/18/060918fa_fact

3. There are also a variety of critiques from other disciplines including philosophy, psychology, and the basic sciences.

4. For a more detailed history, see chapter 1 of Glimcher 2009.

5. It is important to note that the use of the term *rational* in this case is different that the more commonly used definition, which refers to being "logical" or "sane." For economists, *rational* refers to a set of formal axioms that essentially amount to "internally consistent." This is important, as popular media portrayals of NE and BE often do not make this distinction.

6. An important semantic point to note is the use of "neuronal" and not "biological" antecedents of economic behavior. This is to differentiate NE from other research in economics that relates to biology, most notably genetics.

7. Bernheim (2009) provides a detailed discussion of this topic.

8. In the case of variation in economic behavior *between* individuals, NE could prove useful by uncovering neural correlates of patience or other economic traits. This could in theory prove useful to assessing how people might individually respond to a given policy intervention, such as cigarette taxes. Such neuronal markers of behavior are likely years, if not decades away, however.

9. See Heckman (Heckman and Urzua 2010) versus Imbens (Imbens 2010) for one such example.

10. A full discussion of the limitations of fMRI is beyond the scope of this chapter but can be found elsewhere in this volume (e.g., Fitzpatrick). Logothetis (2009)

also provides a relevant, comprehensive, and balanced review of the potential and limitations of fMRI.

11. Ethical issues aside, attempting to maximize neural utility is extremely problematic as the brain is a highly plastic organ that changes in the context of exposure to stimuli, most clearly in the context of addicts developing high tolerance to addictive substances. Thus, any model of neural utility that focuses on direct stimulation would have to account for complex dynamic effects over time.

12. Soma was a drug used to invoke the feeling of pleasure in the inhabitants of London in the distant future, primarily as a way for the government to control the population.

WORKS CITED

Bernheim, B. Douglas. 2009. "On the Potential of Neuroeconomics: A Critical (but Hopeful) Appraisal. Review of 2009." *American Economic Journal: Microeconomics* 1, no. 2: 1–41.

Bernheim, B. Douglas, and Antonio Rangel. 2004. "Addiction and Cue-Triggered Decision Processes. Review of 2004." *American Economic Review* 94, no. 5: 1558–90.

Bonanno, Giacomo, Christian List, Bertil Tungodden, and Peter Vallentyne. 2008. "Introduction to the Special Issue of Economics and Philosophy on Neuroeconomics." *Economics and Philosophy* 24 (Special Issue 03): 301–2.

Camerer, Colin F. 2007. "Neuroeconomics: Using Neuroscience to Make Economic Predictions. Review of 2007." *Economic Journal* 117, no. 519: C26–42.

Camerer, Colin F., George Loewenstein, and Drazen Prelec. 2004. "Neuroeconomics: Why Economics Needs Brains. Review of 2004." *Scandinavian Journal of Economics* 106, no. 3: 555–79.

Camerer, Colin, George Loewenstein, and Drazen Prelec. 2005. "Neuroeconomics: How Neuroscience Can Inform Economics. Review of 2005." *Journal of Economic Literature* 43, no. 1: 9–64.

Camerer, Colin, George Loewenstein, and Drazen Prelec. 2007. "Neuroeconomics: How Neuroscience Can Inform Economics." In *New Developments in Experimental Economics, vol. 2*, edited by E. Carbone and C. Starmer. Elgar Reference Collection. International Library of Critical Writings in Economics, vol. 209. Cheltenham, UK: Elgar.

Choi, James J., David Laibson, and Brigitte C. Madrian. 2004. "Plan Design and 401(k) Savings Outcomes." National Bureau of Economic Research Inc. NBER Working Papers: 10486.

Colander, David. 2007. "Edgeworth's Hedonimeter and the Quest to Measure Utility." *Journal of Economic Perspectives* 21, no. 1: 1–11.

Colander, David, and Andrew Q. L. Chong. 2010. "The Choice Architecture of Choice Architecture: Towards a Non-Paternalistic Nudge Policy." *Journal of Economic Analysis* 1, no. 1: 42–48.

Glimcher, Paul W. 2009. *Neuroeconomics: Decision Making and the Brain*. London: Elsevier Academic Press.

Gul, Faruk, and Wolfgang Pesendorfer. 2009. "A Comment on Bernheim's Appraisal of Neuroeconomics. Review of 2009." *American Economic Journal: Microeconomics* 1, no. 2: 42–47.

Heckman, James J., and Sergio Urzua. 2010. "Comparing iv with Structural Models: What Simple iv Can and Cannot Identify. Review of 2010." *Journal of Econometrics* 156, no. 1: 27–37.

Imbens, Guido W. 2010. "Better Late than Nothing: Some Comments on Deaton (2009) and Heckman and Urzua (2009). Review of 2010." *Journal of Economic Literature* 48, no. 2: 399–423.

Kahneman, Daniel, and Amos Tversky. 1979. "Prospect Theory: An Analysis of Decision under Risk. Review of 1979." *Econometrica* 47, no. 2: 263–91.

Levitt, Steven D., and Stephen J. Dubner. 2005. *Freakonomics: A Rogue Economist Explores the Hidden Side of Everything.* New York: William Morrow.

Lowenstein, George, and Peter Ubel. 2010. "Economics Behaving Badly." *New York Times,* July 14: A31.

McClure, Samuel M., David I. Laibson, George Loewenstein, and Jonathan D. Cohen. 2004. "Separate Neural Systems Value Immediate and Delayed Monetary Rewards." *Science* 306, no. 5695: 503–7.

McClure, Samuel M., David I. Laibson, George Loewenstein, and Jonathan D. Cohen. 2007. "Separate Neural Systems Value Immediate and Delayed Monetary Rewards." In *New Developments in Experimental Economics. vol. 2,* edited by E. Carbone and C. Starmer. Elgar Reference Collection. International Library of Critical Writings in Economics, vol. 209. Cheltenham, UK: Elgar.

Mirrlees, James A. 1971. "An Exploration in the Theory of Optimum Income Taxation." *Review of Economic Studies* 38, no. 114: 175–208.

Rubinstein, Ariel. 2006. Discussion of "Behavioral Economics": "Behavioral Economics" (Colin Camerer) and "Incentives and Self-Control" (Ted O'Donoghue and Matthew Rubin). In *Advances in Economics and Econometrics: Theory and Applications, Ninth World Congress, vol. 2,* edited by R. Blundell, W. K. Newey, and T. Persson. Econometric Society Monographs, no. 42. Cambridge: Cambridge University Press.

Rubinstein, Ariel. 2008. "Comments on Neuroeconomics." *Economics and Philosophy* 24 (Special Issue 03): 485–94.

Thaler, Richard H., and Cass Sunstein. 2008. "Questions for Richard Thaler and Cass Sunstein." Amazon.com Review. Available at http://www.amazon.com/Nudge-Improving-Decisions-Health-Happiness/dp/product-description/0300122233.

Tversky, Amos, and Daniel Kahneman. 2000. "Loss Aversion in Riskless Choice: A Reference-Dependent Model." In *Choices, Values, and Frames,* edited by D. Kahneman and A. Tversky. New York: Cambridge University Press; New York: Russell Sage Foundation.

Ubel, Peter A. 2009. *Free Market Madness: Why Human Nature Is at Odds with Economics—and Why It Matters.* Boston: Harvard Business Press.

Williams, Jonathan, and Eric Taylor. 2006. "The Evolution of Hyperactivity, Impulsivity and Cognitive Diversity." *Journal of The Royal Society Interface* 3, no. 8: 399–413.

Zak, Paul J., and Ahlam Fakhar. 2006. "Neuroactive Hormones and Interpersonal Trust: International Evidence. Review of 2006." *Economics and Human Biology* 4, no. 3: 412–29.

Chapter 10

Functional Brain Imaging

Neuro-Turn or Wrong Turn?

Susan M. Fitzpatrick

Research questions of interest to neuroscientists share a natural overlap with those pursued by scholars studying philosophy, art, music, history, or literature. The common ground is a shared desire to understand the workings of the human mind. What initially attracts someone to study neuroscience, regardless of what aspects of nervous system function an individual career may become focused on (e.g., basic functions of the synapse), is the allure of contributing knowledge that deepens our understanding of our minds. Many neuroscientists want to know how it is that the activities of the cells of the nervous system, individually and collectively, contribute to our ability to see beauty, take pleasure from reading and writing, appreciate music, mull over past experiences, make plans for the future, and contemplate "big questions" such as the purpose of life. Of course these are also questions motivating scholars in the humanities and social sciences. From this shared perspective it is easy to see the appeal the neuroscientific turn holds for those interested in melding collaborations among researchers from neuroscience and scholars from the humanities and social sciences.

This essay does not address what is gained or lost by more closely linking theories, methods, and findings from neuroscience with the theories, methods, or findings in more humanist disciplines. Rather, this essay focuses on a narrower question: do functional neuroimaging methodologies offer a constructive path for the neuroscientific turn in the humanities and social sciences? By reviewing some common misunderstandings and misconceptions about what neuroimaging[1] can and cannot reveal about the mind, particularly concerning BOLD functional Magnetic Resonance

Imaging (BOLD fMRI) or Positron Emission Tomography (PET), I hope to offer one answer to the question.

Although techniques for monitoring brain function by monitoring physiological parameters have been available for more than a century the general appeal of functional brain-imaging techniques as a way of studying the mind[2] is a fairly recent phenomenon. As I will argue later in this essay, I believe the appeal and to some extent the misconceptions of functioning brain-imaging studies is largely due to how the results of such studies have been popularly portrayed both by scientists and journalists.

I have closely followed the development of brain imaging since the late 1970s, first as a scientist in laboratories using both traditional wet-laboratory techniques and PET and later while a member of one of the pioneering in vivo Magnetic Resonance Spectroscopy (MRS) laboratories studying brain metabolism. I initially joined the James S. McDonnell Foundation (JSMF), a philanthropic organization supporting mind/brain research, in 1993 and was surprised at the enthusiasm for and rapid adoption of brain-imaging technologies by cognitive psychologists. Knowing firsthand how difficult it is to bridge the gap from measurements of brain metabolism to inferences about brain function, I thought it premature to use PET and BOLD fMRI to make inferences about complex cognitive and behavioral functions. Little did I suspect that by the end of the 1990s the highly complex physiological data obtained by brain-imaging techniques would morph into the brightly colored "brain scans" described as showing our minds in action and now ubiquitous in newspapers, magazines, and other media. Joe Dumit (who appears in the afterword to this volume) provides an interesting perspective on how functional brain images morphed from a scientific tool for visualizing highly complex neurophysiological clinical and research data to a familiar cultural icon in *Picturing Personhood: Brain Scans and Biomedical Identity* (2003). Dumit's book focuses on studies using PET imaging, but the conceptual and methodological issues he discusses are equally applicable to studies carried out with the now more widely available BOLD fMRI.

Functional brain-imaging studies about the neurological underpinnings of psychopathic behavior, desire, maternal love, the belief in God, why we like luxury goods, and myriad other topics are a natural draw for scientists, scholars, journalists and the general public.[3] But are these topics actually well suited for study by functional brain imaging? I hope this essay will convince readers that the answer to this question is 'no.' Despite their seeming accessibility, functional brain-imaging studies are difficult to design, execute, and interpret. In my experience, functional-imaging studies posing questions about aspects of human cognition that are poorly understood at the psychological and neurobiological levels contribute to general misconceptions about brain and mind and, even more generally, the links between biology and behavior. Too often, the finding that changes in the brain account for some behavior of interest is often pitched as though the brain-behavior link is *the* new, interesting finding. There has been little doubt in either psychology or neuroscience since the nineteenth century that behavior arises from the brain and body. In fact, the role of the brain in behavior was acknowledged at the time of the Hippocratic writings: "From nothing else but from the brain came joys, delights, laughter and jests, and sorrow, griefs, despondency and lamentations" (Plum and Posner 1980, 1).

The results from BOLD fMRI studies are often discussed in the popular media as though differences in observable behavior reveal intrinsic, hardwired differences in brain structure. This common misinterpretation is largely due to a misunderstanding of the capabilities of function brain imaging. Much like the way the results of genomic screens are thought of as more convincing than mere family history (see Marcus 2010), brain scans seem more convincing with respect to how our minds work than psychological data even when the primary nature of the results in a study is behavioral (McCabe and Castel 2007; Weisberg et al. 2008). The idea that brain imaging reveals something previously hidden about the "real mind" has led to a rush to use the widely available technology of BOLD fMRI to "neuro-fy" a number of academic pursuits, resulting in emergent hybrid disciplines including neuroeconomics, neurolaw, neuroethics, neuromarketing, neuroeducation, and neuroaesthetics. What primarily strikes me when reading publications from these attempts is that the hybrids could be more accurately hyphenated with *cognitive* in the place of *neuro-* as the findings rarely reveal novel insights about brain structure-function relationships (for a discussion of this point see van Eijsden et al. 2009) but are usually reinforcing psychological findings.

I will leave it to others in this volume to demonstrate to what degree the knowledge gained by such hybrid pursuits is beneficial, or not, to the ongoing evolution of various academic disciplines. The concern of this essay is how the embrace of functional brain imaging in the development of the emergent neurohybrid disciplines rests on misunderstandings concerning the design and interpretation of functional brain-imaging studies and what information (if any) these visually seductive brain scans actually depict.

A number of thoughtful reviews discuss some of the problems inherent in functional neuroimaging studies in the areas of neuromarketing and social neuroscience, including some discussion of the point I make above, that oftentimes brain imaging is used to merely add "scientific weight" to what is already known from behavioral findings (Ariely and Berns 2010; Cacioppo et al. 2003). If one takes the critical discussions seriously it becomes clear that the primary problems with using functional brain imaging in the "neuroscientific turn" are not technical (although there are technical concerns, to be sure) as much as conceptual. It is not simply a matter of doing such studies better. Most neuroscientific-turn questions are framed assuming that we already understand how neural substrates serve cognitive functions, but these are questions that still need much theoretical and experimental work (for a discussion of this point see Haxby 2010).

My reading of papers from the emergent neurodisciplines suggests we learn very little of substance beyond what was already known from the cognitive psychological or behavioral studies—except for perhaps some converging evidence that behavior is accompanied by brain activity (I return to this point in greater detail below). It is not quite clear to me what other organs in the body were/are hypothesized to be responsible for mental functions. Beyond showing that the brain activity is involved in thinking, deciding, reading, feeling, and so on, what else is gained by resource-intensive brain-imaging studies? One reason given for incorporating functional brain imaging into disciplines such as economics or law is that brain imaging can provide neural *explanations* for complex behaviors. The difficulty of constructing explanations that cross levels of analysis from brain function to cognitive functions to behavior are rarely acknowledged[4] but are central to the difficulties of designing and interpreting brain-imaging studies addressing questions of interest to the humanities and social sciences.

What Does Brain Imaging Image?

Before turning to some examples concerning in what ways brain imaging can and cannot tell us about the mind, it is worthwhile to review the scientific basis of functional brain imaging. Most of the functional brain images we have become accustomed to seeing are provided to us devoid of the experimental details. The lack of details makes it is easy to overlook the technical assumptions and caveats that are required to meaningfully interpret the images. The rapid surge in fMRI technology during the last fifteen years builds on a century of research on brain metabolism and blood flow (Raichle 1998) coupled with a long history of advances in magnetic resonance physics (Logothetis 2008). At one time, most of the individuals using the tools required for functional brain-imaging studies were part of or well-versed in these research traditions. Today, it is more important for newcomers to actively keep in mind that functional brain imaging, be it with BOLD fMRI or PET, does not *directly* measure what is commonly referred to as "brain activity" (Fitzpatrick and Rothman 1999).

Functional brain-imaging techniques rely on the measurement of changes in brain blood flow or metabolism and depend on the observation that brain energy metabolism and neuronal activity are coupled. This coupling is what allows researchers to noninvasively monitor brain function (Roy and Sherrington 1890; Siesjo 1973; Sokoloff 1981). A breakthrough in this storied research effort, and one that made human studies routine, was the development thirty years ago of PET (Raichle 1998). In combination with experimental paradigms and models developed in cognitive psychology, PET allowed the first high-resolution metabolic maps of functionally specialized regions of the human brain. A drawback of the PET technology was its reliance on cyclotron-generated short-lived radioisotopes. The subsequent development of functional magnetic resonance imaging (fMRI), which is also a way to detect changes in brain metabolism and blood flow but does not require the use of radio-labeled compounds, made functional brain mapping widely available (Kwong et al. 1992; Ogawa et al. 1992). It is important to note that in the early days of PET and BOLD fMRI the collaborations typically involved researchers with strong experimental backgrounds in cerebral blood flow and metabolism, neurophysiology, and neuroanatomy working with experts in PET and MR physics. The studies were a direct outgrowth of the traditional attempts to understand the *regional* couplings among blood flow, metabolism, and function. The rise of cognitive neuroscience brought the addition of cognitive psychologists in the 1990s who added expertise in experimental tasks designed to use functional brain-imaging studies to further probe brain structure/function relationships.

Given the current enthusiasm for fMRI in fields like cognitive neuroscience and social neuroscience it is easy to forget that the basis of the signal, the statistical analysis of imaging data, the design of the psychological or behavioral components, and the interpretation of the results remain active and intense areas of research and controversy in the field of cognitive neuroscience even after twenty years of effort. The lessons from cognitive neuroscience should send a note of caution to emerging neurohybrid disciplines.[5]

The detailed technical discussion found below provides a brief introduction to the technical methodology because with functional imaging the devil really is in the details. Prior to crossing too many disciplinary boundaries (remember—functional imaging already combines aspects of neuroscience, biophysics, statistics, and experimental psychology) it is important to have enough local knowledge to ask the right questions. The technical foundations of functional neuroimaging should be kept in mind even if one's only interest is as a casual observer of functional-imaging results. The application of PET and fMRI to localize cognitive processes is based on observations that functional neuronal activity increases when a region is involved in performing a cognitive task (see Posner and Raichle 1994 for a description of imaging study design). These functional neuronal activities are responsible for information processing by neurons and other brain cells and include neurotransmitter metabolism and the generation of neuronal action potentials. The energy required for these and other brain processes is provided primarily by oxidative glucose metabolism (Siesjo 1978). Functional-imaging techniques typically measure glucose metabolism or neurophysiological parameters coupled to glucose metabolism, such as changes in blood flow, changes in regional blood volume, or changes in the rate of oxygen utilization (Sokoloff 1981). The most popular method for performing functional-imaging studies, BOLD fMRI, is sensitive to these physiological parameters but does not directly measure them. BOLD is sensitive to changes in the ratio of oxygenated to deoxygenated hemoglobin in the brain's blood supply (Ogawa et al. 1998). The ratio of oxygenated to deoxygenated hemoglobin is determined by the delivery of oxygen to brain cells and the rate that oxygen is consumed by cellular metabolism.

In a typical functional-imaging study, a subject performs experimental tasks while the signal (the nature of which depends on the technique used) is acquired. The *functional image* depicts the incremental change in signal intensity during a task relative to a baseline state in which the subject rests in the scanner (or relative to a control task). Cognitive processes are localized by functional imaging using experimental paradigms and analyses based upon theories of cognitive neuroscience. As an illustrative example, consider a study designed to assess whether a subject's frontal brain regions are involved in the general cognitive skill of verbal working memory. The subject would perform tasks requiring this cognitive skill, such as remembering lists of words, while being "scanned." In one strategy the degree of involvement of verbal working memory in each task would be varied, but the requirements for other cognitive skills (such as visual recognition) are held constant. The relative intensity of the functional-imaging signal in the frontal region during each task would be statistically correlated with the verbal working memory component (Posner and Raichle 1994).

During an fMRI experiment the "images" obtained, representing the spatial distribution of MR signal intensity changes, are acquired rapidly (on the order of one second each). The large data sets so acquired require the experimenters to use statistical analysis packages to compare the images obtained during different cognitive tasks. The statistical comparison searches for brain regions showing different signal intensities during the different tasks. The final presentation, often called the brain activity map, is usually an image of these statistically identified regions, color coded to represent the level of statistical significance, overlaid on an anatomical MR brain image. From these maps, researchers make inferences about the neural correlates of cognitive processes (for further discussion of this point see Fitzpatrick and Rothman 2002).

If, during conversations with researchers primarily interested in using fMRI as a tool to interrogate cognitive behaviors (and not actively engaged in the technical development of brain-imaging science), I pose the question, "What does the fMRI signal measure?" the answers (in decreasing order of frequency and increasing order of accuracy) tend to be:

regional neuronal activity
incremental changes in regional neuronal activity
incremental changes in regional cerebral blood flow and/or metabolism

As is evident from the technical discussion provided above, none of these descriptions is completely accurate. An MR physicist would describe the most popular fMRI method, blood oxygen level–dependent imaging (BOLD), as measuring the change in the intensity of the nuclear MR signal due to changes in the transverse relaxation time and phase of the protons of water molecules in the blood and brain tissue as a result of changes in hemoglobin oxygenation and blood volume. There are strong links between the BOLD signal and the underlying activity shifts of ensembles of neurons, and these links are active areas of research (Logothetis 2008), but at present none of the functional-imaging methods commonly used directly measures neuronal activity.

The take-home message from the technical details above is that despite the language used to discuss them, the brain images displayed in scientific publications and in the popular media are not representations of changes in brain neuronal activity, or areas of "activation," or the brain "lighting up" or "switching on." Brain scans acquired with fMRI do not even graphically depict the magnitude of the BOLD signal. Rather, the images are computer-generated, color-coded "maps" of statistically significant comparisons among data sets. It must be stressed that the finding of statistically significant differences and a measured change in the actual magnitude of the signal acquired are not necessarily interchangeable. The same BOLD signal may be statistically significant in one subject and not distinguished from noise in another subject. In addition to measurement sensitivity issues, identifying a brain region as having a statistically significant BOLD signal depends on how well the time course of the signal agrees with the assumed time course of the cognitive task. Depending on the nature of the assumptions made in the statistical analysis package, the same data could yield statistical maps identifying different patterns of activation.

Another caveat in directly relating functional images to regional neuronal activity is that even the fiery red spots in the statistical map image overlays from a typical fMRI study may represent changes in neuronal activity of less than 1 percent relative to the brain's regional neuronal activity at rest (Shulman and Rothman 1998; Gusnard and Raichle 2001), and differences between tasks are commonly on the order of 0.1% of the brain's activity when not engaged in an imaging experiment. The realization, largely from other imaging methods such as PET and MRS, that the changes in regional neuronal activity during tasks are quite small has led to the emergence of studies trying to understand the "resting state" through examining changes in the fMRI signal when the subject is not engaging in directed activity (Raichle 2010). It is crucial to note that the lack of a signal change (no "lighting up") does not mean that a region is not involved in the performance of the task studied.

It may be tempting for someone interested in using fMRI to answer questions about the neural correlates of viewing impressionist paintings or to discover what brain regions "light up" when reading Chaucer to dismiss concerns about the missing links in the chain of reasoning from the fMRI signal to neuronal activity and instead concentrate on making inferences about cognition and behavior from patterns of brain activity. However, considering the extent to which the methodology influences the results, it behooves researchers and scholars interested in using fMRI or other brain-imaging modalities to gain a deeper understanding of the theoretical and practical strengths and limitations of the techniques. During a typical BOLD fMRI experiment, for example, over 10 million measurements will be made of which only a small percentage are actually relevant to the experimental question (Haxby 2010; Vul and Kanwisher 2010; Logothetis 2008), and the risk of making interpretation errors is large. Furthermore, even if the imaging methodologies perfectly measured changes in neuronal activity my general concerns about applying functional imaging to particular kinds of questions would still hold. Most important, the critical but often overlooked component of any meaningful fMRI study is that interpretation of changes in neuronal activity is meaningless without a theory of the cognitive operations involved in performing the experimental tasks, and a well designed set of tasks to test the theory. There should also be a testable hypothesis about the neurobiological underpinnings of the cognitive operations under investigation. I expand on the role of understanding the cognitive operations necessary for performing a task in the next section of this essay. In the absence of the framework of a cognitive theory, it is not possible to make inferences about the relationships between an observed behavior and the patterns of brain activity as revealed by functional brain imaging (Poldrack 2008; Henson 2005).

Functional Imaging and Cognitive Neuroscience

The burgeoning interest in the use of functional-imaging studies to interrogate the neural correlates of behavioral and social phenomena is due in part to the role functional imaging played in the development of cognitive neuroscience over the past two decades. The ability to use measurements of brain metabolism and hemodynamic properties to "map" psychological concepts onto physical neurobiological substrates seemed tailor-made for a new academic discipline intending to uncover the neurological underpinnings of cognition (Gazzaniga, Ivry, and Mangun 2002). As already discussed in this essay, the appropriate use of functional brain-imaging methodologies in cognitive neuroscience is still evolving (Haxby 2010; Logothetis 2008) and serves as a cautionary tale for other interdisciplinary endeavors.

A good example of how functional neuroimaging can add to our understanding of the neural substrates of cognition is provided by a now classic paper by the Washington University group known for pioneering applications of brain-imaging methodologies (Petersen et al. 1998). The experimental task design for the study demonstrates how neuroimaging data, when combined with behavioral data and a cognitive theory of task performance, can inform our understanding of the underlying neural systems. The paper summarizes a rather complicated series of experiments investigating a type of learning called procedural learning or skill acquisition. This type of learning is labeled by psychologists as "nondeclarative" in the sense it is difficult for a person to explain or describe what has been learned—although it is clear that performance improves over time. We have all had experience learning a task that seems effortful when new but with practice and time becomes easier and automatic, even though we cannot fully articulate exactly what it is that has been learned. Learning to ride a bicycle or to float in water are familiar examples of how a task feels "different" as we progress from novice to skilled performer.

The Petersen et al. study was motivated by the existence of two possible proposals for how neural modifications could accompany skill acquisition. One explanation was that as a skill became more practiced and expert over time, the neural circuitry involved in the task became more "efficient." The second explanation suggested that as a skill is acquired the nature of the "task" undergoes changes in the way the task is represented and processed in the brain as the novice performance becomes more expert. It was hypothesized that in the "different tasks" case the functional-imaging data would reveal the accompanying changes in neural substrates. Of course, Petersen et al. were quick to point out that it was most likely that these two explanations need not be mutually exclusive and that skill acquisition could involve both mechanisms. In this essay I will only discuss one of the tasks used in the functional neuroimaging studies, a type of maze learning. At first, the performance of an individual engaged in maze learning is slow with many errors. With practice performance speeds up and errors decrease. Together with the behavioral data, the imaging findings supported the idea that as an individual became more expert at the maze task, the relative pattern of activity of different areas of the brain changed consistent with the hypothesis of a change in representation with experience.

The study serves as a good model for, among other characteristics, its intricate and thoughtful experimental design and by today's standards rather modest conclusions. Readers of this essay are encouraged to consult the original paper as it demonstrates how careful one must be when carrying out and interpreting results from a functional-imaging study even for seemingly simple questions about how the brain supports mental processes. Furthermore it built on a strong foundation of prior results. Neuropsychological findings from individuals with discrete brain lesions had previously indicated that procedural learning, unlike learning and memory for "declarative" types of knowledge (e.g., where you went on your last vacation), is not dependent on the medial temporal lobe. Conversely, procedural learning can be impaired by damage to other brain regions, such as the basal ganglia and the cerebellum that have less impact on declarative types of knowledge. It is important to emphasize that the Petersen et al. study was preceded by a robust tradition of studying procedural learning in neuroscience and in cognitive psychology. Generally, functional-imaging data are more likely to advance questions in cognitive neuroscience when the experimental design can call on well-articulated cognitive theories about the nature of the task, solid behavioral data, and neuropsychological evidence about the possible neural underpinnings.

Going Beyond Cognitive Neuroscience

When functioning imaging tools migrate from cognitive neuroscience into other disciplines, even closely related fields such as social neuroscience, the likelihood that the experimental criteria required to meaningfully interpret functional-imaging data can be met is less certain. A recent functional-imaging study, "The Power of Charisma—Perceived Charisma Inhibits the Frontal Executive Network of Believers in Intercessory Prayer," carried out by Uffe Schjoedt and colleagues at Aarhus University in Denmark (Schjoedt et al. 2010) and profiled in a *New Scientist* news piece (Coghlan 2010), provides a good counterexample to the Petersen et al. study. The studies from the Schjoedt lab are, in my view, typical of attempts to neurofy topics of research (social networking, influence) that are traditionally of interest in the social sciences. For the purposes of this essay, Schjoedt et al. 2010 provides an example of the use of fMRI that needs to be pursued with caution. Schjoedt and colleagues compared BOLD fMRI data acquired from self-described devoted Christians (primarily individuals belonging to charismatic Christian denominations) and self-described "secular" participants (the publication states these were mostly BA students) with no prior experience with prayer. The volunteers completed a questionnaire on the strength of their belief in God and the power of prayer prior to the imaging study. After the experiments the participants completed a questionnaire on perceived qualities of the speakers and the feelings they had about God's presence. Brain images were acquired while each participant listened to eighteen different prayers recorded by three different male readers identified to the research subjects as a non-Christian, an ordinary Christian, and a Christian known for healing powers.

The main hypothesis of the paper, that "participants' assumptions about the speaker would affect the evoked BOLD response," was tested by contrasting the scans obtained from the Christians and from the secularists. The secular group showed similar BOLD responses across the three conditions in the study. Analysis of the scan acquired from the Christians "only revealed significant activations in one of the contrasts, namely, in the contrast 'non-Christian' relative to 'Christian known for his healing powers.'" The main finding reported was the inhibition of a frontal executive network in the Christian participants. The paper speculates the "inhibition" (technically a decrease in the measured BOLD signal relative to that found in the secular participants) contributes to the facilitation of charismatic influence.

Like the Petersen et al. study described above, the Schjoedt et al. paper is rather detailed, and I recommend that interested readers consult the original publication. Unlike the Petersen et al. study, which made an important contribution to our understanding of the neural mechanisms underpinning procedural learning, it is not quite clear, at least to me, what we learn from the Schjoedt et al. study. In light of all the caveats I have discussed that can alter the findings in a functional-imaging study, what new contributions to our understanding of behavior, cognition, or neural function does this study provide us? We do not have a well-characterized behavioral or cognitive theory of charisma as we do for learning as measured by improved performance on the maze task and the other tasks used in the Petersen et al. study. There is not, to my knowledge, a testable hypothesis, based on cognitive science, neuropsychological studies, or findings from patients with brain injuries that could suggest how "charisma" is represented in the brain. The findings of the Petersen et al. study also sound a warning bell that should be heeded for any functional-imaging study—experience changes the cognitive representation and the neural underpinnings of the task. The self-selected, self-identifying devout Christian subjects willingly admitted to strong faith and a deep belief in the power of intercessory prayer, while the secular participants denied such beliefs. Listening to prayers being read is, at its most fundamental, a different behavioral and cognitive task for each of these subject groups. For Christians praying is familiar (practiced), salient, and rewarding, and it may evoke an array of rich associations. Prayers by other Christians can have a power that words recited by an identified nonbeliever may not. For non-Christians much of the rich context for prayer is missing. Is it surprising to see different patterns of BOLD responses as the two subject groups listened to prayer? What the study cannot tell us is *why* certain individuals become devout Christians and others do not, despite the report of a fundamental difference between the brain scan patterns of the two groups.

The Schjoedt et al. study does not consider a common confound with functional-imaging studies comparing two groups selected for their differences—it is not possible to control for experience, beliefs, prior knowledge, attentional effects, affect, or the many other factors we encounter in life that influence cognition and behavior. The criticisms I am making here are not dependent on the technical and methodological difficulties discussed above. To me, the Schoedjt et al. studies are conceptually problematic; they derive from misguided attempts to use BOLD fMRI to directly map mind onto brain. The most, following the Petersen et al. model, we can take away from the Schjoedt et al. paper is that listening to prayers is a different task for devout Christians compared with non-Christians. This is not surprising, as the social science data told us as much.[6]

Separating the Wheat from the Chaff

In 1998, the James S. McDonnell Foundation (JSMF) began hosting the Neuro-Journalism Mill.[7] This site, which brings together popular stories and the original functional-imaging studies that inspired them, reveals some of the conceptual and methodological problems that arise when attempting to use fMRI to probe notions like charisma, or courage, or romantic love. Both the scientific studies and the media reports suggest that brain-imaging studies reveal the innate, biological roots of behavior in a way that *explains* why humans act the way they do. Of concern is that such studies arise from views about the capabilities of fMRI, or for that matter how the brain works, that may be more in line with portrayals for popular consumption rather than with the more limited uses dictated by the constraints of neuroscience, experimental cognitive psychology, physics, and imaging statistics.

We should be on our guard against falling for "the brain as metaphor," a tendency to equate parts of the brain with the attributes of the tasks invoked during a particular study. Analysis of imaging studies involving higher cognitive functions is likely to show that a brain region is involved in any number of tasks that share common cognitive attributes (such as attention) or similar kinds of computations. Labeling brain areas or regions as "fear detectors" or "emotion meters" or "conflict monitors" creates a misconception that there is a one-to-one correspondence between a function and a brain "part" (Poldrack 2008). Another difficulty arises when a part of the brain is equally active in both the experimental and the control conditions, which (as described above) can even include an individual resting quietly in the scanner. Although the level of activity does not change, the contributions of the cells in that area of the brain can still be essential for carrying out the task; however, as most functional brain images depict "differences" (via subtraction), the contribution of a particular region of the brain to a behavior of interest could be masked. Even in the more complex and reflexive efforts to understand the cognitive and emotional processes underlying complex behaviors, an important limitation is that functional brain-imaging experiments are constrained by a number of important characteristics: the subject is lying down and inaccessible inside the scanner. The tasks used to elicit a particular pattern of brain activity are often experimental proxies standing in for complex behaviors that, in everyday life, have real consequences (Ariely and Berns 2010).

The Cognitive Paparazzi

The idea that with functional brain imaging we can visually access the inner workings of our minds holds powerful allure for anyone interested in human nature. The clichéd adages that "a picture is worth a thousand words" and "seeing is believing" are to some extent based on the fact that humans, like other primates, have brains well adapted for interpreting visual information (see Chalupa and Werner 2003). It is for this reason that photojournalists, recognizing the power of visual images, have strict ethical standards regarding the use of photographs,[8] knowing that images convey strong messages that may not necessarily fully represent the context under which they were obtained.

To recognize and analyze the seductive power of fMRI data visualization, the James S. McDonnell Foundation (JSMF) organized a session for the 2005 annual meeting of the American Association for the Advancement of Science (AAAS).[9] The session, titled "Functional Brain Imaging and the 'Cognitive Paparazzi': Viewing Snapshots of Mental Life Out of Context," was inspired by the work of the paparazzi: photographers who make a living by aggressively obtaining candid photographs of celebrities—often by violating their privacy. The photographs, splashed across the pages of celebrity tabloids, are presented to us freed of any situational context except that provided by the accompanying cutline. A picture of a movie star looking unkempt for whatever reason can appear on the front page of a tabloid headlined "X Distraught over Betrayal by Y." By endlessly dogging celebrities, the paparazzi attempt to reveal what lies behind the carefully presented public image—the "real person." Similarly, the cognitive paparazzi want to use functional brain imaging to peer candidly into the privacy of our "minds" and reveal something about behavior that would otherwise remain hidden. And, like the tabloid pictures, once brain scans are in the public domain, the brightly colored images are also freed from the controlled experimental context during which they were obtained.

One question explored during the session, and one I still find puzzling, is, why are the findings presented in brain scans considered to be accessible and readily interpretable by nonexperts? Most popular science articles generally provide some abstracted version of the main findings of a study but do not provide the actual scientific data presented in the original scientific publications. Brain scans seem to be an exception. I hope readers of this essay are convinced that functional brain-imaging scans are highly technical and difficult to interpret without expert knowledge of the subjects participating in the studies, the tasks performed, the techniques used to acquire the data, and the complicated statistical tools used to analyze the data and create the images. Philosopher Adina Roskies (2010) argues that brain scans, like photographs, shorten the inferential distance. Although tempting, we cannot draw conclusions from looking at functional brain images without the full scientific context, a point reinforced by several leading MR labs in a recent perspective piece suggesting guidelines for reporting methods and results for fMRI studies (Poldrack et al. 2008). If functional-imaging data presented in scientific papers for specialized readers require extensive experimental detail for adequate interpretation, and even then uncertainties and questions remain, should not the images presented at multidisciplinary meetings or to the general public also be accompanied by the scientific context and caveats required to support the claims being made?

To conclude, let me give my answer to the question I ask in the title of this essay. Attempting the "neuro-turn" via functional neuroimaging is most likely a wrong turn. Scholars considering what it is that functional brain imaging can bring to their fields should start by asking themselves whether they have a testable hypothesis constrained by cognitive theories and behavioral data that can be brought to the scanner or whether they risk falling into one or more of the traps described in this essay.

Now, my negative answer to this essay's title pertains only to the use of brain imaging and does not mean that there are not questions of mutual interest that are already providing and could potentially provide fertile soil for collaborative interests among neuroscientists, cognitive psychologists, and scholars in the humanities. As the Hanson and Bunzl (2010) volume attests, functional imaging studies are benefiting from collaborations among cognitive psychologists, mathematicians, and philosophers. Researchers are beginning to ask if the cognitive processes involved in the formation and recall of individual memories may play a role in what is called collective or group memory (Roediger et al. 2009). A recent paper in the *Proceedings of the National Academy of Sciences* (Hughes et al. 2010) demonstrates how mathematical tools developed by vision scientists can be used to distinguish between the paintings of Pieter Bruegel the Elder and imitators. In a *Nature* "News and Views" two theoretical neuroscientists discuss the Hughes et al. study and suggest a number of ways statistical image analysis tools could be helpful to art historians (Olshauser and DeWeese 2010). In an event hosted jointly by the St. Louis Academy of Science and a St. Louis arts organization, CraftAlliance, philosopher Mark Rollins explored how research in the cognitive sciences is altering our interpretation of how perception may impact the development of cultural styles in artistic expressions.[10] Interesting opportunities are created at disciplinary boundaries as, for example, when anthropologic, cognitive psychological, and psychiatric views of mental dysfunction are thoughtfully considered (Boyer 2010).

I suspect scientists will have their ideas about human cognition and behavior enriched by working closely with scholars who observe human endeavors through the varied lenses of different academic traditions. The structure and function of the brain is determined both by biology and by our experiences. The collaborative efforts that occur because of natural shared interests that shape and define the questions being pursued together with a shared commitment to respect and understand the strengths and limitations of borrowed concepts and tools are likely, in my view, to have the best chance for fruitfully contributing to our collective knowledge.

NOTES

1. In this essay, discussion of functional neuroimaging, functional brain imaging, and brain scans primarily refers to studies and images involving Blood-Oxygen-Level-Dependent Magnetic Resonance Imaging (commonly referred to as BOLD MRI or fMRI), or in some cases to studies monitoring cerebral blood flow or metabolism with Positron Emission Tomography (PET). Neuroscientists have at their disposal an increasingly diverse array of brain-imaging and brain-sensing technologies including a number of new tools based on specialized applications of MR and PET physics. BOLD fMRI is the technique most likely to be used for studies involving tasks of interest to scholars in the humanities. PET and fMRI are typically used in the studies appearing in the popular press.

2. I want to emphasize that in cognitive neuroscience and social neuroscience the use of the term *the mind* is not strictly defined and is often used to refer to some amalgam of brain, cognition, and behavior.

3. A number of websites post links to and provide commentary on popular media articles based on neuroscience research. See, for example, "The Neurocritic" (available at http://neurocritic.blogspot.com) or "The Neuro-Journalism Mill" (available at http://www.jsmf.org/neuromill/about.htm).

4. A philosophical analysis of the difficulties of constructing such multilevel explanations in the neurosciences can be found in *Explaining the Brain* by Carl Craver (Oxford University Press, 2007) and *Mental Mechanisms* by William Bechtel (Lawrence Erlbaum Associates, 2007).

5. An edited volume, derived in part from work related to a collaborative activity grant made to Steven Hanson and Clark Glymour by JSMF, examining the

fundamental theoretical assumptions about neural organization and cognitive function that have underpinned the use of functional neuroimaging in cognitive neuroscience studies and how new theories and mathematical tools are causing the field to revisit some of the assumptions, has recently been published (see Hanson and Bunzl 2010) and provides an excellent introduction to the methodological and conceptual issues related to functional brain-imaging studies.

6. For additional analyses of problematic functional-imaging studies of relevance to the "neuroscientific turn" see Loosemore and Harley 2010.

7. The site, originally called Bad Neurojournalism, was later renamed the Neuro-Journalism Mill and taglined as "separating the wheat from the chaff of popular writings about the brain." The Mill continued the JSMF tradition of critiquing "bad neurojournalism" but also highlighted examples of popular science writing "getting it right." The Mill was shuttered in 2009, but an archive is maintained at http://www.jsmf.org/neuromill/about.htm.

8. See, for example, the National Press Photographers Association Code of Ethics (available at http://www.nppa.org).

9. A summary of the session is available at http://www.jsmf.org/neuromill/cognitive-paparazzi2.htm.

10. For details see http://www.craftalliance.org/news/art&science7–10.htm.

WORKS CITED

Ariely, D., and G. S. Berns. 2010. "Neuromarketing: The Hope and Hype of Neuroimaging in Business." *Nature Reviews Neuroscience* 11, no. 4: 284–92.

Boyer, P. 2011. "Intuitive Expectations and the Detection of Mental Disorder: A Cognitive Background to Folk-Psychiatries." *Philosophical Psychology* 24, no. 1: 95–118.

Cacioppo, J. T., G. G. Berntson, T. S. Lorig, C. J. Norris, E. Rickett, and H. Nusbaum. 2003. "Just Because You're Imaging the Brain Doesn't Mean You Can Stop Using Your Head: A Primer and Set of First Principles." *Journal of Personality and Social Psychology* 85, no. 4: 650–61.

Chalupa, L. M., and J. S. Werner. 2003. *The Visual Neurosciences.* Cambridge: MIT University Press.

Coghlan, A. 2010. "Brain Shuts Off in Response to Prayer." *New Scientist*, April 24, 9.

Dumit, J. 2003. *Picturing Personhood: Brain Scans and Biomedical Identity.* Princeton: Princeton University Press.

Fitzpatrick, S. M., and D. L. Rothman. 1999. "New Approaches to Functional Neuroenergetics." *Journal of Cognitive Neuroscience* 11:467–71.

Fitzpatrick, S. M., and D. L. Rothman. 2002. "Meeting Report: Choosing the Right MR Tools for the Job." *Journal of Cognitive Neuroscience* 14:806–15.

Gazzaniga, M. S., R. B. Ivry, and G. R. Mangun. 2002. *Cognitive Neuroscience: The Biology of the Mind.* 2nd ed. New York: W. W. Norton.

Gusnard, D. A., and M. E. Raichle. 2001. "Searching for a Baseline: Functional Imaging and the Resting Human Brain." *Nature Review Neuroscience* 2:685–94.

Hanson, S. J., and M. Bunzl. 2010. *Foundational Issues in Human Brain Mapping.* Cambridge: MIT Press.

Haxby, J. V. 2010. "Multivariate Pattern Analysis of fMRI Data: High-Dimensional Spaces for Neural and Cognitive Representations." In *Foundational Issues in Human Brain Mapping,* edited by S. J. Hanson and M. Bunzl, 55–68. Cambridge: MIT Press.

Henson, R. 2005. "What Can Functional Neuroimaging Tell the Experimental Psychologist?" *Quarterly Journal of Experimental Psychology* 58A, no. 2: 193–233.

Hughes, J. M., D. J. Graham, and D. N. Rockmore. 2010. "Quantification of Artistic Style through Sparse Coding Analysis in the Drawings of Pieter Bruegel the Elder." *Proceedings of the National Academy of Sciences* 107:1279–83.

Kwong, K. K., J. W. Belliveau, D. A. Chesler, I. E. Goldberg, R. M. Weisskopf, B. P. Poncelet, D. N. Kennedy, B. E. Hoppel, M. S. Cohen, R. Turner, H. M. Cheng, T. J. Brady, and B. R. Rosen. 1992. "Dynamic Magnetic Resonance Imaging of Human Brain Activity during Primary Sensory Stimulation." *Proceedings of the National Academy of Sciences,* USA 89:5675–79.

Logothetis, N. K. 2008. "What We Can Do and What We Cannot Do with fMRI." *Nature* 453:869–78.

Loosemore, R., and T. Harley. 2010. "Brains and Minds: On the Usefulness of Localization Data to Cognitive Psychology." In *Foundational Issues in Human Brain Mapping,* ed. S. J. Hanson and M. Bunzl, 217–40. Cambridge: MIT Press.

Marcus, A. D. 2010. "How Genetic Testing May Spot Disease Risk." *Wall Street Journal,* May 4.

McCabe, D. P., and A. D. Castel. 2007. "Seeing Is Believing: The Effect of Brain Images on Judgments of Scientific Reasoning." *Cognition* 107:343–52.

Ogawa, S., R. S. Menon, S. G. Kim, and K. Ugurbil. 1998. "On the Characteristics of Functional Magnetic Resonance Imaging of the Brain." *Annual Review of Biophysics and Biomolecular Structure* 27:447–74.

Ogawa, S., D. W. Tank, R. Menon, J. M. Ellermann, S. G. Kim, H. Merkle, and K. Ugurbil. 1992. "Intrinsic Signal Changes Accompanying Sensory Stimulation: Functional Brain Mapping with Magnetic Resonance Imaging." *Proceedings of the National Academy of Sciences,* USA 89:5951–55.

Olshausen, B. A., and M. R. DeWeese. 2010. "The Statistics of Style." *Nature* 463:1027–28.

Petersen, S. E., H. van Meir, J. A. Fiez, and M. E. Raichle. 1998. "The Effects of Practice on the Functional Anatomy of TaskPerformance." *Proceedings of the National Academy of Sciences* 95, no. 3: 853–60.

Plum, Fred, and Jerome B. Posner. 1980. *The Diagnosis of Stupor and Coma.* 3rd ed. Philadelphia: F. A. Davis.

Poldrack, R. A. 2008. "The Role of fMRI in Cognitive Neuroscience: Where Do We Stand?" *Current Opinion in Neurobiology* 18:223–27.

Poldrack, R. A., P. C. Fletcher, R. N. Henson, K. J. Worsley, M. Brett, and T. E. Nichols. 2008. "Guidelines for Reporting an fMRI Study." *NeuroImage* 40:409–14.

Posner, M. I., and M. E. Raichle. 1994. *Images of Mind.* New York: Freeman Press.

Raichle, M. E. 1998. "Behind the Scenes of Functional Brain Imaging: A Historical and Physiological Perspective." *Proceedings of the National Academy of Sciences* 95, no. 3: 765–72.

Raichle, M. E. 2010. "Two Views of Brain Function." *Trends in Cognitive Science* 14, no. 4: 180–90.

Roediger, H. L., F. M. Zaromb, and A. B. Butler. 2009. "The Role of Repeated Retrieval in Shaping Collective Memory." In *Memory in Mind and Culture*, edited by P. Boyer and J. V. Wertsch, 139–70. Cambridge: Cambridge University Press.

Roskies, A. L. 2010. "Neuroimaging and Inferential Distance: The Perils of Pictures." In *Foundational Issues in Human Brain Mapping*, edited by S. J. Hanson and M. Bunzl, 195–215. Cambridge: MIT Press.

Roy, C. S., & C. S. Sherrington. 1890. "On the Regulation of the Blood Supply of the Rat Brain." *Journal of Physiology* (London) 11:85–108.

Schjoedt, U., H. Stødkilde-Jørgensen, A. W. Geertz, and A. Roepstorff. 2009. "Highly Religious Participants Recruit Areas of Social Cognition in Personal Prayer." *Social Cognitive and Affective Neuroscience* 4, no. 2: 199–207.

Shulman, R. G., and D. L. Rothman. 1998. "Interpreting Functional Imaging Studies in Terms of Neurotransmitter Cycling." *Proceedings of the National Academy of Sciences, USA* 95:11993–98.

Siesjo, B. K. 1978. *Brain Energy Metabolism*. New York: Wiley.

Sokoloff, L. 1981. "Relationships among Local Functional Activity, Energy Metabolism, and Blood Flow in the Central Nervous System." *Federation Proceedings* 40:2311–16.

van Eijsden, P., F. Hyder, D. L. Rothman, and R. G. Shulman. 2008. "Neurophysiology of Functional Imaging." *NeuroImage* 45:1047–54.

Vul, E., and N. Kanwisher. 2010. "Begging the Question: The Nonindependence Error in fMRI Data Analysis." In *Foundational Issues in Human Brain Mapping*, edited by S. J. Hanson and M. Bunzl. Cambridge: MIT Press.

Weisberg, D. S., F. C. Keil, J. Goodstein, E. Rawson, and J. R. Gray. 2008. "The Seductive Allure of Neuroscience Explanations." *Journal of Cognitive Neuroscience* 20, no. 3: 470–77.

Chapter 11

A Clinical Neuroscientist Looks Neuroskeptically at Neuroethics in the Neuroworld

Peter J. Whitehouse

As a clinically oriented cognitive neuroscientist, I celebrated the past successes and future promises of science during The Decade of the Brain (1990–2000). I listened as we were told that we were well on our way to understanding how the mind worked, how we might enhance our thinking, and how the use of stem cells and other powerful biological approaches would eventually cure neurological diseases such as Alzheimer's. I heard the call for a new field of ethical inquiry to help us manage these powerful new abilities to manipulate ourselves and others.

My participation in the party was, however, with halfhearted, limited to hemibrained enthusiasm. The triumphalism of Western, left-hemispheric scientific approaches was loud and clear, but it seemed as though it was missing the big picture. As we entered the new millennium, the hype about malleability of the brain manifested in the concept of neuroplasticity oozed into social and mental spaces in a way that seemed first stretched and then rigid. Now that we knew that the brain could change, everything else apparently could now be altered. The prefix *neuro-* began to be applied widely to intellectual and creative pursuits, even to the word *society* itself (Restak 2006). Somehow, understanding the brain appeared to be our salvation from the ennui associated with the post-Freudian sense of the apparently unfathomable imbroglio of the psyche. Had the neuro-logical completely displaced the psycho-emotional as the defining feature of our humanity?

As psychologist Ken Gergen (2010) has pointed out, the brain is best viewed as the organ that has evolved through time to support our work and play as cultural beings. However, we seem to focus on the inverse, as the objectives and methods of science have somehow progressed with the goal of understanding the brain and its associated molecules and to dissolve the relevance of the psychological and the social, and perhaps with them, our very humanity. Historians such as Anne Harrington (1992) contextualize conversations about the brain in broader issues at work in society (for example, the tendencies to think of brain function in parts or as a whole). Fernando Vidal (2009) coined the term *brainhood* to point out the ahistorical, political, cultural, and ethical aspects of the effort to replace personhood with a purely neurological concept. His idea supports those of us embroiled in the day-to-day practices of treating and researching conditions that affect the brain. How do we enhance person-centered care and rescue our patient/client from those who wish to rush him or her off to the scanner before even engaging in conversation? Do we really think that the self is nothing more than the brain, or that the brain is synonymous with the mind? Vidal also strongly supports the position argued in this chapter that a co-opted neuroethics (and bioethics) can be more harmful than helpful by attracting our attention to less important issues (e.g., enhancing thinking with drugs) and away from more critical ones (improving understanding about global climate issues). Perhaps we need to replace these ethical frameworks with a broader social discussion about the profound implications of a Medicine gone wrong.

Here is our road map for this endeavor. First, I build on these psychological and historical critiques of the current brain-obsessed movement launched by the Decade of the Brain. I reflect skeptically on the claims of neuroscience to this position of power over our humanity by focusing particularly on the emergence of the field of neuroethics and more specifically on the scientific, social, and ethical responses to the challenges of brain aging, such as so-called Alzheimer's disease (AD). The primary focus of neuroethics has been on the alleged power of science to explain and modify human behavior. Yet, as an ambitious subfield of bioethics, it shares with its ideological progenitor a lack of historical perspective and concern for the future beyond the agenda of science and medicine. I argue that a narrowly focused neuroethics is not contributing as it could to some of the ethical issues we currently face.

In conclusion, I will suggest how we might begin to create more integrative and wiser intergenerational narratives and more empowering learning spaces in communities, a perspective that I hope will broaden the scope of contemporary neuroethics and strengthen its force as a critical perspective on the neuroscientific turn. By rethinking aging-associated cognitive challenges such as dementia, and understanding that broad social responses are more important than narrow medical ones, we may actually see the cognitive limitations and problems with activities of daily living in a new and more profound light. A reframed bioethics true to its origins and a reimagined neuroethics more aligned with the needs of the future could help, rather than hinder, such an essential effort (Potter 1971; Whitehouse 2003b).

Personal Perspective

As a clinical neurologist with training in psychiatry, a PhD, a fellowship-trained cognitive neuroscientist, and a master's level prepared bioethicist, I have been turning in and around the "neuro" for my entire career. As an MD-PhD student at the Johns Hopkins University I studied the cognitive challenges of persons with schizophrenia and then of those with stroke. Several fairly constant themes emerged for the remainder on my career: the cognitive and biological aspects of attention, verbal and visual memory, and interhemispheric differences (Whitehouse 1981). As a junior faculty member, I became (for a short time perhaps) the world's expert on a single brain nucleus (the variably named Substantia Innominata, which is now referred to as the Cholinergic Basal Forebrain) involved in the neuropathology of so-called Alzheimer's disease (Whitehouse 1982). I also mapped neurotransmitter receptor proteins in the autopsy specimens from human beings with a variety of disease conditions (Whitehouse 1988), which contributed to imaging them in life, and published one paper on that topic (LaFrance 1981). At Case Western Reserve University I founded an Alzheimer's Center (later renamed Memory and Aging Center) that attempted to build bridges between the "hard" neurosciences and the "softer" psychological and social sciences and between research and care. First concerned about the ethical issues in the care of patients with dementia and then the ethical issues of bioethics itself, I then obtained formal training in bioethics. As one of the first in the era of modern neuroscience writing on the ethics of cognitive enhancement and of dementia (Whitehouse et al. 2000), I was well positioned to be a founding member of the neuroethics movement but chose not to become involved.

During my career, I walked back and forth between the (often indistinct) political and social lines that separate and/or join the brain and the mind. For example, I was elected to membership in the American College of Neuropsychopharmacology and later cofounded the International College of Geriatric Psychoneuropharmacology. Whether the *psycho-* precedes the *neuro-* in the compound word or vice versa relates to both marketing and politics. On occasion, I observed leading academic psychiatrists becoming quite paranoid about losing neuroscience as their basic science to the neurologist's exclusive provision. For example, as the neurologist director of an NIMH Clinical Research Center on Psychopathology of the Elderly, I attended academic and advocacy meetings in Washington where the clear message to each other and to Congress was that the brain belongs to psychiatry too. I was interested in psychiatry because it was allegedly the most biopsychosocial of the medical fields. But as the academic capital of the "neuro" increased, the psychiatrists themselves were running away from their cultural roots toward the lofty inspiring heights of the various branches of neuroscience.

I have conducted clinical studies, planned and participated in a few neuroimaging efforts, designed drug trials, and employed a variety of other quantitative and qualitative approaches to understand the relationship between brain function and cognition. Despite the many claims that we have made tremendous strides in understanding these relationships between biology and psychology, I think we have no (good) idea about how the mind and brain connect. Part of the marketing of science is to claim an ever-increasing number of breakthroughs as part of what is portrayed as a never-ending march to progress. I have faith in science as an endeavor that has contributed to human well-being, but scientism—an unbridled and unquestioning faith in science—seems to me dangerous and unwise.

By this stage of my career I had become a neuro-skeptic. My tongue-in-cheek approach illuminates my skepticism about using the *neuro-* prefix to enhance the credibility of the various terms it modifies. I emphasize my usually gentle sarcasm by pointing out that my skepticism is fully informed by the latest neuroscience (now where are those centers of skepticism and sarcasm networks in the brain?). As the neurologist and ethicist Eran Klein (2009) has pointed out, others have deployed the neuro-logism "neuro-skepticism" in their quest to understand the turn to neurophilia, if not neuro-obsession. The philosophically and clinically trained Klein considers various depths of neuroskepticism from a clinical perspective. In other words, just how much does neurological knowledge inform the care of different kinds of patients? He also reminds us that John Dewey felt skepticism was characteristic of a healthy mind (particularly a scientifically oriented one).

My skepticism has varying degrees of intensity depending on the word following the prefix *neuro-*. The more general and less science-based the word, the more likely it has little relevance to direct brain modification.

Neurosociety seems the grandest; *neuroscience* seems the most legitimate given the way that the different disciplines of neuroanatomy, neurophysiology, and neuropharmacology reorganized themselves several decades ago to form a single enterprise. *Neuropsychology* seems likewise legitimate as an enterprise assessing the impact of brain disease on cognition and behavior. *Neurolaw, neuropolitics, neuroeducation, neuromarketing, neuroaesthetics, neurotheology,* and *neuroleadership* seem questionable whenever a claim is made that brain science will reconfigure these fields of human activity to a significant degree. *Neurophilosophy* has been around a while, but its subfield of *neuroethics* is relatively new. Perhaps, as we will see, the invention of neuroethics may even be dangerous. Currently it attracts our attention toward seductive, allegedly powerful, medical tools and away from its own constructs of risk and risk management. Except in discussions with my patients, who probably prefer a professional physician to an amateur, I claim amateur status in many of these fields and am tempted at times to refer to myself as a generalist, an "ologist" if you will, that is, even eliminating the prefix from my primary professional designation. I eschew being labeled a bioethicist (and even dropped my secondary appointment in the then newly formed Department of Bioethics) especially since it is unclear whether bioethics is a field, a discipline, a profession, or, as I came to believe, a misguided, shortsighted enterprise.

Bioethics and Its Caricaturist Child, Neuroethics

Bioethics as a field and now an emerging profession is only a few decades old. Yet, to a considerable degree, it has displaced the humanities and arts in medical education and perhaps even in the practice of value exploration in the health field. For example, rather than having educators from the humanities teaching medical students about the broader, moral context of medicine, bioethicists, often ensconced in medical schools, dominate the teaching roles. Bioethicists are perhaps less likely to challenge the reductionist, scientistic, and profit-motivated goals of medicine. They are often dependent on funding from local sources (like the medical school dean), government agencies (e.g., ELSI [Ethical, Legal, and Social Issues program of National Human Genome Research Institute]), and even the medical industry. The latest subfield of bioethics to emerge is neuroethics, which was given its label by William Safire (2002a, 2002b) as a field to explore the ethical implications of specific medical neurotechnologies. Others, such as Michael Gazzaniga (2005), still focus on the implications of neuroscience for society but argue for a much broader scope. But what would it mean to have a brain-based philosophy of life, as he asks us to consider? Neuroethics appears to be largely concerned only about the value implications of new or imagined technologies, and it is principally populated by those who espouse an optimistic view of potent and imminent neuro-progress.

The history of bioethics itself is relevant as well since neuroethics aspires to be a subfield of the parent discipline. Van Potter, who coined the term *bioethics*, initially proposed the field as a kind of environmental land ethic, a position he built atop the work of American ecologist Aldo Leopold (Potter 1971). In a time of global climate change our responsibilities for preserving the integrity, stability, and beauty of the land (Leopold 1948) seem all the more essential thirty-five years later. Should it not be a neuroethical issue that we still allow the brains of our children to be exposed to known neurotoxins such as lead, mercury, and other poisons? Where are the neuroethical activists, and should more advocacy of this nature be expected of them? Bioethics can serve as a "bridge to the future" (the title of Potter's first book) and should be a fundamentally intergenerational and interconnecting ethic. We can no longer just uncritically promote the imagined power of overly simplistic scientific solutions to the complex challenges of human and natural system failures.

We must attend to enhancing our cognitive capacities to address these challenges. Yet as the number of older people increase throughout the world, dementia—and specifically so-called Alzheimer's disease (AD)—threatens these very capacities. Medicine is struggling with understanding the nature of dementia and the nosology of classes of cognitive impairment (Fotuih et al. 2009), as illustrated by the imprecision and uncertainly associated with the term *Alzheimer's disease*. Dementia is the broad category that encompasses all conditions that cause generalized cognitive impairment in someone who was previously intellectually intact and is not affected by a medical illness causing temporary confusion. Alzheimer's disease is said to be the most common cause of dementia, but it is itself a broad category with many indistinct boundaries between it, other dementias, and aging itself.

As a clinician researcher who cares for those with memory problems who are at risk for "Alzheimer's" and who also studies the more general domain of cognitive enhancement, I entered the world of ethics through concerns about diagnostic disclosure, conflict of interest, informed consent, genetic testing, and end-of-life care (Kodish 1996; Karlawish 2000; Whitehouse and Sachs 1996; Whitehouse and Juengst 1997; Post and Whitehouse 1995; Post and Whitehouse 1998a and 1998b; Whitehouse 1998). I have become aware of the limitations of our neuro-language (imprecise labels, for example) and the dangers of pursuing a kind of neuro-utopia through promised cures. The fearful concept of "Alzheimer's" has for some time been uncritically assumed to be a single condition clearly differentiable from "normal aging." Rather, many who have studied the entity "Alzheimer's" for decades are coming now to view it as multiple heterogeneous processes lumped together under a hundred-year-old eponym (Whitehouse, Ballenger, and Mauer 2000; Whitehouse and George 2008). These conditions overlap with aging and other dementias and are not likely to yield to magic bullet, molecular fixes. Yes, many accept this broad term without much deep consideration of the utility of this social construction. Labels like *Alzheimer's disease* create fear and stigma, and rob individuals of the opportunity to create more positive stories of aging, which can incorporate decline and eventual death in different ways than the model conception of an incurable fatal disease. And yet, elders with cognitive impairment can develop the personal strength to continue contributing to their communities despite their short-term memory problems.

The Image of the Changeable Neuroworld

I will make the critique of neuroethics more concretely and the neuroscientific turn more generally using my own viewpoint as a clinical neuroscience researcher. The use of neuroimaging as a tool to understand the brain and then the use of drugs to improve brain function will be our targets. The emergence of the neuroscientific turn is illustrated by the Positron Emission Tomography (PET) image, which is marketed as having the power to reveal all the secrets of thought and illness. If the neuroimage is the face of the neuroscientific turn, then neuroplasticity is certainly its underlying central dogma. Now we can "change the brain" so the world can be fixed, or so the message goes. Yet my skepticism goes deeper and asks whether there is a role for what might be called *psychoplasticity* in this fabric of ideas celebrating our human behavioral adaptability. What matters more: to change our brain or to change our thinking, attitudes, and behavior? Obviously the two kinds of changes go hand in hand, but what is our ultimate goal, and which approaches are more likely to achieve it?

Neuroimaging, particularly PET scans and fMRI, became the poster child for the notion, increasingly common during the Decade of the Brain, that neuroscience was the most exciting field of human enterprise. How many covers of public and professional magazines of all kinds have been graced by a multicolored oval-shaped image that is claimed to represent the brain at work? How many times have we seen two comparison images, one "normal" and the other a specific disease state that is somehow explained by these two images? The differences are not only claimed to be black and white but made more persuasive by being displayed in multicolored pictures. As Racine (Racine et al. 2005; Racine and Zimmerman, this volume) demonstrates and captures with the term *neurorealism*, dramatic images make the concept being discussed seem more salient and somehow obvious. Surely we need more studies of the psychological impact of such images and how their claim to objectivity (Dumit 2003) can be challenged. As Vidal has correctly suggested, perhaps the artistic imagination should be brought to bear on the fanciful and fantastical images produced by numbers from a scanner. Pseudocolored renditions of brain function appear to be scientific renditions of facts not open to interpretation, but what they really require is a critical and an aesthetic hermeneutic.

Neuroethics could help us understand the rhetorical force of this imaging technology by asking particular questions: How have epistemological traditions encouraged scientists and journalists to display diametrically opposed images to make their point? Are they hiding (the often) messier data in tables in the body of the article? How many of these images have been enhanced, and to what end, by using careful selection of colors to make differences more apparent? How many times have data been averaged across subjects to show greater consistency than actually exists? How often have neuroimagers been frank about the challenges of reliability associated with their techniques? The answers to these interrelated questions are not certain. However, I know from personal experience and from anthropological studies (Dumit 2003) that many such distortions have occurred and are occurring. For example, the National Institute of Mental Health (NIMH) had a PET scan on their website long after the specific image had been discredited in the scientific literature.

The science behind these images is often more problematic than one might think. What some have called the new phrenology of bright colorful spots on images (rather than bumps on the head) has real and fundamental limitations. As many neuroscientists move toward thinking of the brain as systems of networks the neuroimagers are to a large degree stuck in the brain centers model (Uttal 2003). Do laypeople who observe these images appreciate that the blood flow or glucose utilization represents some average of activity of inhibition and excitation? That in a particular cognitive task it might be the uncolored dark, apparently silent areas, of the brain that are integrating information from across the brain or are really more responsible for the cognitive or emotional functions being studied? Our marketing of neuroimaging findings is often too full of exaggeration given the scientific limitations discussed above.

Neuroplasticity—The Way to Global Salvation?

Let me now turn to an essential concept of the Decade of the Brain: neuroplasticity. This is said to be the foundation of the definitive therapies. The discovery that nerve cells can grow and sprout in the adult brain is claimed to be the revolutionary concept that will transform human life. We now know that the brain can change! Is this a surprise? Does our knowledge of how it changes help us therapeutically? The brain responds to input and hence is said to change form like plastic. We are not the victims of hardwiring, so the claim goes, but our brain can be reprogrammed after illness or other events. Perhaps words like *plasticity* (a wonderful metaphor derived from material science) and *programming* (from computer science) were chosen to emphasize that science can work wonders in changing the form and function of even an organ as complex as the brain.

There is no doubt: the science is interesting. We know new connections can be made (but this is not such new information) and even new neurons can appear (this is newer information and exciting at a basic discovery level). This information is said to offer huge promise for enhancing our brain function and treating those with neurological conditions.

Yet what educator, psychologist, or critical thinker across the ages would not have imagined that the brain could change? After all, we learn, and we recover from neurological insult. Perhaps when we thought that the liver or some other abdominal organ was responsible for thinking and the brain merely a radiator to cool the blood we might not have focused so much on brain changes. However, once we learned that the brain was the principal organ of thought, we must have appreciated that change in cognition and behavior means a change in the brain.

Did we ever celebrate psychoplasticity? Now we might refer to brain-savvy teachers as neuroeducators, but previously perhaps they were psychoplastic teachers. The history of psychology, particularly in the new era of cognitive neuroscience, shows us that the quest for biological substrates trumped the quest for the social foundation of thinking. As mentioned earlier, Ken Gergen has pointed out that the brain ought to be viewed as the primary organ of enculturation. Hence we need to turn the neuro around and have a socio-turn. After all, the brain does learn and help us act in a cultural space defined by how we apply names to our bodies and organs.

But so too the concept of "brain" itself is a cultural construct (see Kismet Bell, this volume). Certainly components of the brain such as the so-called limbic system (meaning limbus or ring of structures) are labeled crudely in order to try to understand it as a "system" that plays a role in emotions. Even the somewhat appropriately labeled almond-shaped amygdala is composed on multiple subnuclei and is tightly linked to other parts of the brain, not a solitary center of feelings and passion. After all we think with our body, not just our brain (e.g., Wilson 1998), as other nonneural systems such as hormones clearly influence our mental life. It is probably best also to recognize we think socially, and even when we are alone our memories of others influence our own thinking profoundly. We are socially cognitive beings, not brains in vats.

Now admittedly, understanding the mechanisms by which the brain changes does offer some promise. It is not exactly clear how it would affect, say, classroom instruction, but it might lead to the development of new biological therapies, be they drugs or cell implants that might improve the lot of human beings. However, despite many claims to the contrary the concept of neural plasticity has not led to any effective interventions as yet. Drugs and other biological products such as stem cells have not created brain rewiring, and it might be that psychological interventions may in fact modify brain plasticity better than drugs. Educational interventions including information technology can have profound effects on the learner; still, we must be cautious about claims for any restricted computer-based technology to enhance brain function more effectively than a good teacher and a supportive school.

The Broader Cultural Context of Medicine and Society

The issues of neuro-hype clearly relate to the challenges of scientism in our society as a whole. Our brain seems to have a connection with our self, and clearly it is a relevant organ in contemplating what makes us human. Yet the claims that neuroscience is accelerating rapidly into a brave new world of neural enhancement is part of a more general problem: our cultural belief that science can solve every exigent challenge.

Arguably, medicine is excessively enamored of new technologies and new sources of profit. The U.S. health care system is not sustainable financially, and this situation is partly driven by the widespread and early adoption of unproven diagnostic and therapeutic approaches. One clear illustration of this is the emergence of genomic medicine and its claim to be the foundation of the future of the much-heralded personalized medicine. Perhaps the gene is the only thing that can compete with the brain as a biological element that lays some claim to our essential humanity. Indeed, genetic medicine suffers from some of the same conceits as neuromedicine.

The current director of the National Institute of Health, Francis Collins, is an exuberant, proselytizing genomicist (Collins 2006) whose vision of the future is one in which the primary care physician has genetic material of each individual patient on his office computer. This vision of medicine's future rarely acknowledges the myriad complexities of using genetic information to help patients and potential patients. My own work with the genes for ApoE, a cholesterol transport protein that has been implicated as a factor in increasing risk for Alzheimer's disease, demonstrates some of these complexities. It is difficult to calculate individual risks for diseases because persons are affected by age, ethnic background, and other factors. The models to calculate risk are based on limited data sets that may not represent the same gene pools of the individuals seeking help. Moreover, ApoE is pleotropic, meaning that it affects risk for a variety of different health conditions involving the brain, heart, and perhaps other organs including the eyes. Hence, giving accurate risk information is very difficult today despite the hype. Can human beings incorporate all this complex information, even if the patient's physician (1) understands the genetic information and how it relates to the patient's health status and (2) has spent adequate time with the patient to map personal values to medical options? In fact, the field of genomics is beginning to recognize that its theoretical potential has not been realized. For example, Genome Wide Association Studies have not led to as much understanding about disease risk as expected, and a bit of humility may be necessary.

The Example of Alzheimer's Disease

As has been described throughout this chapter, so-called Alzheimer's disease represents a good example of the dangers of what we might think of as the neuro*scientistic* turn. In fact, Alzheimer's disease combines the fascination and power of both genes and brains. The myth of Alzheimer's is that it is one disease; again, most experts see it as several biological processes converging on a variable picture of progressive cognitive impairment. At every level of description—genetic, pathological, and clinical—there is considerable heterogeneity. Whether this variability warrants referring to this as Alzheimer's diseases in the plural or a syndrome is an important matter.

Practically, it means that single biological approaches based on molecular reductionist approaches are likely to be limited. Cures seem an empty but frequently made promise.

The primary motivator for the claim that molecules can generate substantial diagnostic and therapeutic benefits is genetic medicine. Early therapies were based on an understanding of systems neuroscience, that is, which nerve cells are damaged and how their communication with other neurons using neurotransmitters is impaired. The nucleus I studied is characterized by cell loss in so-called Alzheimer's disease and other dementias. The neurons of this nucleus uses the neurotransmitter acetylcholine. Current drugs are quite limited in their effects; for example, they prevent the breakdown of acetylcholine and allow the neurotransmitter to function longer. However, these drugs provide only very modest symptomatic benefit in some people and do not alter the biological progression.

However, once genes were discovered that caused Alzheimer's disease in rare families and seemed to act through the processing of amyloid-related proteins, the stakes were raised by the scientists who claimed that they could now prevent or cure the disease by disrupting the basic biology of amyloid processing. Now the promise was not to just treat symptoms but to slow biological degeneration. There are many problems with these claims, and the amyloid hypothesis is slowly falling out of favor as an exclusive answer to Alzheimer's disease. As a result of clinical and pathological variability, there are no clear biomarkers that qualitatively discriminate Alzheimer's disease from normal aging. The many confusing efforts to establish and label states between normal and dementia are also being challenged. Recent efforts to promote biomarkers (neuroimaging and cerebrospinal fluid tests) as diagnostic tests are far ahead of our diagnostic and therapeutic knowledge. To me these efforts are somewhat desperate efforts to create a picture of progress without actually demonstrating benefit to people with cognitive impairment or society at large. Mild cognitive impairment (MCI), now divided into early and late forms, is thought to represent some state intermediate between normality and Alzheimer's disease. The use of the term *MCI* only highlights that we are labeling people along a continuum of a variety of age-related processes. Is the greatest challenge to the development of molecular therapies that we are, at the end of the day, seeking the fountain of youth for the brain?

So this disease or set of diseases or illnesses or conditions of severe brain aging represent an example of the neuroscientific turn. We cannot seem to get it out of our mind that this phenomenology of age-related cognitive challenge is most fundamentally a two-word eponym, and also a concept about a chronic degenerative brain disease that can change. This is why I sometimes say that AD is more a disease of the mind than of the brain and why the biologization of Alzheimer's disease is part of the neuroscientific turn. Brain changes sound basic and foundational, but perhaps changing words and stories (which, after all, involve brain changes) is ultimately more important.

Changing the Story

Perhaps we need to consider reevaluating the way we think and talk about some brain diseases. This should be a fundamental issue for a broader neuroethics. If we want to learn how to live a brain-based philosophy of life, as Gazzaniga claims that we should (2005), then we should think deeply about (and value) human conditions that we label as brain diseases. Nor should we privilege the neuroscientists as somehow more important contributors to discussions about ethics and the brain. Although the anecdote is considered the weakest form of evidence in science and the randomized control trial the strongest, in many areas of life the story trumps data.

If I'm in conversation with someone describing these broader aspects of human cognition, I often point out that at that moment in the conversation we are thinking together. My thoughts are not independent from the person with whom I'm speaking or whom I'm reading about or listening to. There is something intrinsically social about cognition that gets to its fundamental nature, its evolution as a human capability. Since we are engaged in changing the story of brain health, I celebrate the relatively recent reemergence of narrative medicine and narrative ethics (e.g., Charon 2006). Medicine is enamored with molecules. But as Muriel Rukeyser is often quoted as saying, "the universe is made up of stories not atoms." Human beings begin by labeling parts of the world be they be a small atom or a large universe. Words taken in isolation, but particularly when combined into stories, can have profound effects on our behavior.

As a former brain banker (the person responsible for arranging donations, collecting tissues, storing them, and making them available for research) at the Johns Hopkins University, I was pleased to develop a new concept of a bank that might contribute to health, namely, a StoryBank (Whitehouse and George 2009). We are collecting a corpus of stories starting with those that relate to people's appreciation and knowledge of genes and moving to stories of volunteering in their communities. Rather than biological scientists crowding around the autopsy table dissecting and sharing the brain specimens, we imagine historians, cognitive scientists, philosophers, English professors, and social scientists examining a story from their different perspectives. Stories represent an opportunity not to dissect our brains and lives into component parts but to put them together in some kind of integrated fashion. Narrative is one basis for the creation of wisdom.

Wisdom

Synthetic or integrated knowledge will be as important as or more important than expertise and narrow disciplinary understanding of the many challenges we now face as embodied brains in an increasingly global community. I have suggested we need to think about the concept of wisdom in modern terms and understand its relationships to quality of life (Whitehouse 2001) and also about its biological substrates and how they relate to attentional mechanisms and cognitive enhancement (Whitehouse 2003a). We need serious discussion about the limitations and enhancements of our thinking and valuing processes. It is not clear how much the brain sciences will contribute to this endeavor. It seems quite possible that focusing on the neuroscience of brain changes may distract us from the major job of rethinking and revaluing our dominant cultural beliefs and actions.

In my own career, I have moved from trying to understand the brain and behavior of a thirsty rat to the ecological and social justice issues associated with an increasingly thirsty human species, including the creation of an innovative public charter school that serves the needs of youngsters, adults, and elders (George and Whitehouse 2010; George and Singer 2010). As the supply of clean water continues to diminish, the temperatures increase, and the ocean waters rise, the epidemiology of dementia will change. Children will have infectious diseases, nutritional deficiency, and trauma-induced cognitive impairment, and, for many reasons, such as environmental threats or social neglect, elders will be at risk for dying sooner. These are important ethical issues being ignored for the most part by bioethics and neuroethicists.

Never have we needed to celebrate more the powers of our human minds and imaginations to promote our flourishing. Never has the organization of communities of thinkers and doers been more important. So why is it that a group of people believe that the alleged imminent power of science to understand the brain will contribute so much to useful knowledge? Can we eliminate terrorism by detecting it in a brain scan? Can we find a pill that treats our inability to think for the long term? Will a drug cure the anxieties and depression associated with the collapse of our communities?

No, they will not, in my view. The quest for magic solutions will distract us from the hard social work ahead. I find hope in our collective humanity and in the humility that comes with recognizing our limitations, not in biological silver bullets. Neuroscience is an exciting adventure that I have had the pleasure to participate in and contribute to, but we should turn away from the obsession with the power of science and the lack of moral imagination that can occur as a result. Only an integrated and deeply contextualized neuroethics has the wisdom to lead the way in the emerging neuroworld.

WORKS CITED

Abram, David. 2010. *Becoming Animal: An Earthly Cosmology.* New York: Pantheon.
Charon, Rita. 2006. *Narrative Medicine: Honoring the Stories of Illness.* New York: Oxford University Press.
Cohen, C. A., P. J. Whitehouse, S. G. Post, S. Gauthier, A. Eberhart, and L. LeDuc. 1999. "Ethical Issues in Alzheimer Disease: The Experience of a National Alzheimer Society Task Force." *Alzheimer's Disease and Associated Disorders* 13, no. 2: 66–70.
Collins, Francis. 2006. *The Language of God: A Scientist Presents Evidence for Belief.* Tampa, FL: Free Press.
Dumit, Joseph. 2003. *Picturing Personhood: Brain Scans and Biomedical Identity.* Princeton: Princeton University Press.
Elliot, Carl. 2010. *White Coat, Black Hat: Adventures on the Dark Side of Medicine.* Boston: Beacon Press.
Fitzpatrick, S. "Functional Brain Imaging: Neuro-turn or Wrong Turn?" This volume.
Fotuhi, M., V. Hachinski, and P. J. Whitehouse. 2009. "Changing Perspectives regarding Late Life Dementia." *Nature Reviews Neurology* 5, no. 12 (December): 649–58.
Gazzaniga, Michael S. 2005. *The Ethical Brain.* New York: Dana Press.
George, D., and P. Whitehouse. 2010. "Intergeneration Volunteering and Quality of Life for Persons with Mild-to-Moderate Dementia: Results from a 5-month Intervention Study in the United States." *Journal of American Geriatric Society* 58, no. 4: 796–97.
Gergen, K. 2010. "The Acculturated Brain." *Theory and Psychology.* 20, no. 6: 795–816.
Harrington, Anne. 1992. *So Human a Brain.* Boston: Birkhauser.
Illes, J. "Preface: A Neuro-Pivot." This volume.
Karlawish, J., P. J. Whitehouse, and R. H. McShane. 2004. "Silence Science: The Problem of Not Reporting Negative Trials." *Alzheimer's Disease and Associated Disorders Journal* 18, no. 4: 18–182.
Klein, E. 2009. "To ELSI or not to ELSI Neuroscience: Lessons for Neuroethics from the Human Genome Project." *AJOB Neuroscience Journal.* doi: 10.1080/21507740.2010.510821.

Kleinman, Arthur. 1988. *The Illness Narrative: Suffering, Healing, and the Human Condition.* New York: Basic Books.

Kodish, E., T. Murray, and P. J. Whitehouse. 1996. "Conflict of Interest in University-Industry Research Relationships: Realities, Politics, and Values." *Academy of Medicine* 71, no. 12: 1287–90.

LaFrance, N. D., H. N. Wagner Jr., P. J. Whitehouse, E. Corley, and T. Duelfer. 1981. "Decreased Accumulation of Isopropyl-iodoamphetamine (I-123) in Brain Tumors." *Journal of Nuclear Medicine* 22:1081–83.

Leopold, A. 1948. *The Sand County Almanac.* New York: Oxford University Press.

McGilchrist, Iain. 2009. *The Master and His Emissary: The Divided Brain and the Making of the Western World.* New Haven: Yale University Press

McKibben, Bill. 2010. *Earth: Making a Life on a Tough New Planet.* New York: Times Books.

Post, S. G., and P. J. Whitehouse. 1995. "Fairhill Guidelines on Ethics of the Care of People with Alzheimer's Disease: A Clinical Summary." *Journal of the American Geriatric Society* 43:1423–29.

Post, S. G., and P. J. Whitehouse. 1998a. "Emerging Antidementia Drugs: A Preliminary Ethical View." *Journal of the American Geriatric Society* 46:1–4.

Post, S. G., and P. J. Whitehouse, eds. 1998b. *Genetic Testing for Alzheimer's Disease: Ethical and Clinical Issues.* Baltimore: Johns Hopkins University Press.

Post, S. G., and P. J. Whitehouse. 1998c. "The Moral Basis for Limiting Treatment: Hospice and Advanced Progressive Dementia." In *Hospice Care for Patients with Advanced Progressive Dementia,* edited by L. Volicer and A. Hurley, 117–31. New York: Springer.

Potter, Van Rensselaer. 1971. *Bioethics: Bridge to the Future.* Englewood Cliffs, NJ: Prentice-Hall.

Potter, V. R., and P. J. Whitehouse. 1998. "Deep and Global Bioethics for a Livable Third Millennium." *The Scientist* 12:1:9.

Racine, E., O. Bar-Ilan, et al. 2005. "fMRI in the Public Eye." *Nature Reviews, Neuroscience* 6 (February): 159–64.

Restak, Richard. 2006. *The Naked Brain: How the Emerging Neurosociety Is Changing How We Live, Work, and Love.* Chatsworth, CA: Harmony Press.

Safire, William. 2002a. "Neuroethics belongs in Public Eye." *Dayton Daily News,* May 17, 13A.

Safire, William. 2002b. "Visions for a New Field of Neuroethics." Paper read at Neuroethics: Mapping the Field, at San Francisco.

Senge, Peter, et al. 2008. *The Necessary Revolution: Working Together to Create a Sustainable World.* New York: Doubleday.

Uttal, William. 2003. *Distributed Neural Systems: Beyond the New Phrenology.* New York: Sloan Educational Publishing.

Vidal, F. 2009. "Brainhood, Anthropological Figure of Modernity." *History of the Human Sciences* 22:5.

Whitehouse, P. J. 1981. "Imagery and Verbal Memory Encoding in Left and Right Hemisphere Damaged Patients." *Brain Language* 14:315–32.

Whitehouse, P. J. 1999. "The Ecomedical Disconnection Syndrome." *Hastings Center Report* 29, no. 1: 41–44.

Whitehouse, P. J. 2003a. "Paying Attention to Acetylcholine: The Key to Cognitive

Enhancement." *Progress in Brain Research*. The Netherlands: Elsevier Science B.V.

Whitehouse, P. J. 2003b. "The Rebirth of Bioethics: Extending the Original Formulations of Van Rensselaer Potter." *American Journal Bioethics* 3, no. 4: W26–W31.

Whitehouse, P. J. 2010. "Taking Brain Health to a Deeper and Broader Level." *Neurological Institute Journal* (Spring): 17–22.

Whitehouse, P. J., J. Ballenger, and S. Katz. 2001. "How We Think (deeply but with limits) about Quality of Life: The Necessity of Wisdom for Aging." *International Library of Ethics, Law and the New Medicine*, vol. 12, *Aging: Decisions at the End of Life*, edited by D. N. Weisstub, D. C. Thomasma, S. Gauthier, and G. F. Tomossy, 1–19. Amsterdam, The Netherlands: Kluwer Academic Press.

Whitehouse, P. J., and D. George. 2008. *The Myth of Alzheimer's Disease: What You Aren't Being Told about Today's Most Dreaded Diagnosis*. New York: St. Martin's.

Whitehouse, P. J., and D. George. 2009. "The Art of Personalized Medicine— 'Banking' on Stories for Healthier Cognitive Aging." *Lancet* (April) 4: 373.

Whitehouse, P. J., E. T. Juengst, M. Mehlman, and T. Murray. 1997. "Enhancing Cognition in the Intellectually Intact: Possibilities and Pitfalls." *Hastings Center Report* 3:14–22.

Whitehouse, P. J., and C. R. Marling. 2000. "Human Enhancement Uses of Biotechnology: The Ethics of Cognitive Enhancement." In *Encyclopedia of Ethical, Legal and Policy Issues in Biotechnology*, vol. 2, edited by T. H. Murray and M. J. Mehlman, 485–91. New York: John Wiley & Sons.

Whitehouse, P. J., Andrea M. Martino, Kendall A. Marcus, Richard M. Zweig, Harvey S. Singer, Donald L. Price, and Kenneth J. Kellar. 1988. "Reductions in Acetylcholine and Nicotine Binding in Several Degenerative Diseases." *Archives Neurology* 45:722–24.

Whitehouse, P. J., K. Maurer, and J. F. Ballenger. 1998. "The Concept of Alzheimer's Disease: Past, Present, and Future (Meeting Report)." *Neuroscience News* 1, no. 4: 39–40.

Whitehouse, P. J., S. G. Post, and G. A. Sachs. 1996. "Dementia Care at the End of Life: Empirical Research and International Collaboration." *Alzheimer Disease and Associated Disorders* 10, no. 1: 3–4.

Whitehouse, P. J., D. L. Price, R. G. Struble, A. W. Clark, J. T. Coyle, and M. R. DeLong. 1982. "Alzheimer's Disease and Senile Dementia: Loss of Neurons in the Basal Forebrain." *Science* 215:1237–39.

Wilson, Elizabeth. 1998. *Neural Geographies: Feminism and the Micro-Structure of Cognition*. New York: Routledge.

Chapter 12

The Mind-Sciences in a Literature Classroom

Bruce Michelson

Neuroscience is fast developing the technical and conceptual where-
withal to reveal in fine, bare detail the neurobiological substrates of
the mind. Perhaps it will despoil a sacred myth—the myth of selfhood
and souls. And if so, we may be wandering innocently into the opening
phase of a dangerous game. Our ethics and systems of justice, our entire
moral order, are founded on a notion of society as a collective of individual selves—autonomous, introspective, accountable agents. If this
self-reflective, moral agent is revealed to be illusory, what then? (Broks
2003, 48–49)

The controversy is severe; provocative research is evolving rapidly—and
for the humanities the stakes are higher than usual. To the study of cultural
history and the teaching and criticism of imaginative literature, this revolution under way in the neurosciences is not just another chance to play with
terminology and imperfectly understood perspectives borrowed from other
disciplines. With potential to destabilize and transform so many common-
place assumptions about the nature of thinking and the dynamics of consciousness, twenty-first-century neuroscience challenges basic conventions
about the structure and continuity of the mind and the self, conventions
that subtend not only everyday humanist discourse about "interpretation,"
"meaning" and "art," but also foundational ideas of who and what we really
are. That's big. Moreover, all of this has to be dodged or negotiated in a
season when cultural artifacts and practices we have produced and contemplated for centuries—the codex of paper to read and scribble about; the
stable physical image and artifact to contemplate; the biohazardous, inter-

personal classroom and seminar where we earn our professional keep—are undergoing changes and challenges more fundamental than any in the previous six hundred years. Compelled to reimagine the mind, we are also obliged to help reconceive the technology and epistemology of what we study and to determine what we can plausibly provide to a culture in the throes of unprecedented adventure.

As all of this unfolds, responsible humanists will have plenty to keep them busy. Even so, as we drift into these crises, it would be nice if we could tell ourselves that the profession's record in coming to terms with such disruptions has been graceful. No such luck. When bandwagons have rolled past our doorway from other precincts of research, we have been known to scramble out and play along without really knowing the music. Luckily, academic practice has grown so specialized and fragmented that we can normally get away with being out of tune, since nobody else is listening. If Alan Sokal's disruption of *Social Text* a few years ago is academic folklore now, it was also a fluke—not the complacent incompetence that his hoax unmasked, but the public embarrassment that it triggered. In the home department, or down in the catacomb meeting-rooms at convention hotels, our Shakespeareans can do riffs on Marx and Negri safe from hostiles with street-cred in Economics; liberties with Lacan can be taken without facing stray hecklers from Psychology; generic English profs can second-guess Salman Rushdie's cosmopolitan ethos with zero risk that Rushdie will find out or care. Engagement with contemporary neuroscience, however, can plunge a humanist into contemplations and responsibilities as challenging and consequential as any that the guild has ever faced. And more people will be watching.

Also the music this time is hard, in terms of the sheer complexity of what we would borrow or merely seek to understand. Theoretical texts and constructs that have caught the attention of scholastic humanists since 1970 have been, for the most part, logical or imaginative rearrangements of accessible historical record or aesthetic experience, arguments configured from the outset for an intelligent laity, and not grounded in esoteric research. This time, the brute fact is that serious and direct engagement with high-tech contemporary neuroscience takes most humanist scholars way out of their depth. Nonetheless, if this is a big part of the cultural moment in which we live, then we must find a way to bear witness and participate from perspectives founded in our own expertise. If we cannot be involved, and do so as something better than amateur intruders, we are headed for trouble—and if we aren't careful, obsolescence.

But involved how? Tempting as it might be to indulge in bust-outs of omnidisciplinarity, this looks like a moment for prudent self-awareness and at least a teaspoon of humility, as we try to sort this out. In rough times all over for American academe, with campus administrators hunting for units and operations to trim back or sacrifice, there is no advantage to bolstering impressions that programs in the crosshairs have no clear or consequential subject, no coalesced methodology, or a penchant for promulgating nonsense. How then might we keep in range of what we actually know, and refresh and empower the kinds of inquiry that have made the humanities valuable and special?

A place to begin: evolving from the study of classics, philosophy, and philology, pursuits that until the later nineteenth century constituted the center of the British and American college curriculum, disciplines in the modern humanities have focused scholarly attention and classroom inquiry upon informed and creative response—within the text under scrutiny as a document informed by cultural and historical circumstance; and also response *to* the text, as aesthetic and intellectual experience for a modern audience. With that recognition in mind, two questions at the outset might assist a conversation about where we are now and what is going on, a conversation to support and validate, among other things, a course or seminar worthy of a full semester of student time and effort, opening and deepening an inquiry that could engage thoughtful scholars and citizens long after the academic term is over.

> About the nature of the mind, consciousness, and the self, what kinds of documents are in play in the culture now, beyond the sites and circles of the actual research?
>
> Over the past century, what texts of enduring interest might provide a not-so-distant mirror for our own predicament? In other words, what breakthroughs and radical redescriptions with regard to the mind have resonated in modern cultural history before now—and where can effects be seen in literary and cultural practice, as those changes have caught on?

In framing answers to either of these questions, we should concede that problems lurk in the available descriptive language. Of necessity, other writers in this volume have argued for categorical distinctions between the *translation* of neuroscientific research and the *popularizing* of it, recognizing that each term does come to us smirched with unhelpful connotations. Because it signifies discourse faithful to meanings of words and phrases in a different language, *translation* cannot gracefully encompass informed and provocative inquiry into social, moral, or political implications of new knowledge; *popularizing*, as the recourse term, may suggest opportunism, reduction, and kitsch, along with unintended hints of condescension on the part of the classifier, given that so few people beyond the labs are qualified to second-guess descriptions or explanations of new work in these disciplines. Reaching a worldwide audience early in the twentieth century, Whitehead's *Science and the Modern World* was more interesting than a translation or a popularizing of recent work in mathematics and physics; Piaget's clear and profound observations in *The Psychology of the Child* are still in a different league from mass-market parent advice at the mall bookstore; and from renowned and active neuroscientists, bold and speculative books like Damasio's *Looking for Spinoza*, Broks's *Into the Silent Land*, or Libet's *Mind Time* do not belong in a stack with Lehrer's *Proust Was a Neuroscientist*, Gladwell's *Blink*, or other such commentaries written from the bleachers.

Also with regard to categories and word-choice, one other adjustment can be useful if historical perspective is a hope. Though *neuroscience* may encompass many varieties of systematic inquiry into the nature and dynamics of the brain, the landscape of the mind, and the nature of the self, connotations of the word are inevitably rooted in our moment, in this powerful recent convergence of fMRI, AI, pharmacology, physical and organic chemistry, systems biology, and other disciplines. When we push back farther for perspective, we encounter influential documents that constituted the best work (or in some cases the most notorious) in their respective historical periods, pronouncements with staying power and cultural consequences, and models of consciousness that continue to subtend critical thinking and imaginative literature in our own time. But Hegel, alas, had no laboratory; Marx could support his radical construction of consciousness with nothing resembling empirical science or reliable statistics; and though Nordau and Lombroso seem ludicrous or contemptible now with their head-calipers and their abusive raids on Darwin, there they are nonetheless, kicking up plenty of whitewater in the cultural record. Understanding what they all did, and what influence they achieved, can strengthen our own speculations about the cultural and moral consequences of the neuroscience coming at us now, genuine, popularized, or butchered. To gather and talk about all this material, a looser term like *mind-science* may do the job better.

For a sustained inquiry in a humanist tradition, and for a good course about the cultural impact of new conceptions of consciousness and thought, we have an abundance of ambitious material to look at: from actual researchers, extending back to the Gilded Age, there are works with literary aspirations and polish; there are also important texts by avowedly literary folk—again, more than a century's worth—engaging and feeling the effects of this turbulence; and in our own moment, gaudy action in popular culture. Embarking from Hegel and Coleridge, and moving forward through Marx, Pater, William James, and Freud, the lineage is rich—and because students in many of our academic disciplines know little or nothing about it, they cannot read, with anything like a full measure of understanding and enjoyment, novels and short stories by major authors who try to countenance these challenging new formulations about what a sentient human being really is. Luckily, we have in our own time a strengthening array of exceptionally well-written texts suggesting an exciting convergence. Publishing for audiences beyond their peers, several high-profile researchers in the neurosciences have also opened a dialogue with literary traditions and strategies. These writers have been "crossing the Quad," as it were—not only to promulgate their own research but also to imagine implications and consequences for understanding the human condition. They take formal and stylistic risks; they meditate on cultural history; they echo and reify classic voices. Because that kind of experiment may eventually constitute a new and powerful genre, it deserves attention from cultural scholars and teachers.

Moreover, in this cluster of recent texts, one recurring theme stands out, especially for the perplexed in our own guild: several of these writers favor a new description for the individual self, a metaphor that has been noted and commented upon elsewhere in this collection. A twenty-first-century neural self may be emerging now as an *authoring* process, as a compulsory and unending labor of inscription and revision, a life-narrative in constant development, edition, transition. If the Cartesian adage has needed a reboot, here it comes: not "We *think*," but "We *revise*, therefore we are." In humanities corners on the Quad, as aging descriptions of the self as a linguistic construct, or as a political or economic construct (bequests of long-dead theorists who could know nothing of modern neuroscience) are being swept aside by new paradigms, this proffered model of the self as a *literary* construct could just possibly fill the void—or to put it another way, a case for the continuing relevance of literary and cultural studies might be salvaged with help from a least-expected source.

Other readings to bring into such a course may be self-evident, as there are many works of fiction out there that resonate with these shifts and upheavals in how we imagine the dynamics and consequentiality of thought and the mind. Henry James, Kate Chopin, Charlotte Perkins Gilman, D. H.

Lawrence, William Gibson, Sebastian Faulks, Dan Lloyd, Richard Powers: though this is a provocative and challenging theme in modern fiction—and potentially one of the most important themes we can explore under these current circumstances—a tour through curricula and course descriptions offered in humanities departments will suggest that such perspectives are still unusual as classroom subjects. And when we move to questions about how contemporary neuroscience echoes in American popular culture, avalanches of material are never more than a few mouse-clicks away. Online media news and blog rumor about mind-science; films and television series informed or contorted by AI possibilities and dreams or nightmares about the programming, overhaul, or replication of the mind, personality, soul—ample fuel for freewheeling exchanges, late in a lively semester, on how all of this is affecting not only our entertainment industries but also the high-circulation myths of a global public.

To sum up: a plan for a course in mind-science and its impact on modern culture can move, more or less cogently, from charted territories of cultural history into disorders and unknowns of the present. To negotiate the past without getting mired in it, a collection of excerpts can do the job—selections from Hegel, Coleridge, Marx, Schlegel, Pater, and others, to achieve historical perspectives appropriate to a single-semester course, and also to interest students in reading further in these texts and authors, perhaps as part of a major writing project. Because some of the most significant conflicts in American cultural history, with regard to new findings, doctrines, and dogmas related to the mind, unfold at the end of the nineteenth century, I also recommend including in this set Elizabeth Cady Stanton's "The Solitude of Self," a surprising and nearly forgotten public address to a U.S. House of Representatives committee near the end of Stanton's career; other good choices are the famous, appalling preface from George Beard's *American Nervousness*; and from right around the same moment, Charlotte Perkins Gilman's short story "The Yellow Wall-Paper." To sample some of the most influential writing from right now, other brief excursions are useful: excerpts from Pinker's *The Blank Slate*, Dennett's *Consciousness Explained*, Donald's *A Mind So Rare*, Gazzaniga's *Human*; and perhaps as dessert, a couple of brief jeremiads by Jean Baudrillard on the postmodern condition, the catastrophe of AI and the triumph of the virtual, and how everything in our *monde* is going straight to the *chiens*. It probably goes without saying that these readings can be spiced on the fly by linking, from a course website, to ephemeral news items, blogs, and online commentaries. The longer works in the course plan are listed below in an order that makes historical and thematic sense.

William James, *Psychology: The Briefer Course*
Henry James, *The Turn of the Screw and Other Short Stories*
Kate Chopin, *The Awakening*
D. H. Lawrence, *Studies in Classic American Literature*
Paul Broks, *Into the Silent Land: Travels in Neuropsychology*
Antonio Damasio, *Looking for Spinoza: Joy, Sorrow, and the Feeling Brain*
William Gibson, *All Tomorrow's Parties*
Dan Lloyd, *Radiant Cool*
Richard Powers, *The Echo Maker*

To keep the inquiry moving forward and provide a measure of continuity, a few questions can be posed at the outset and returned to along the way.

When scientific or philosophical descriptions of consciousness, thinking, and the self draw the attention of writers and artists, where can we see the impact?

How might our awareness of prevailing or subversive descriptions of personality, thinking, and the mind enrich or complicate our understanding of literary works from the same era in the American past?

When a seductive ideology about the nature and the structure of the mind eventually loses preeminence or credibility, what are, and what should be, the consequences for understanding and responding to art produced under the influence of that thinking?

When twenty-first-century researchers in the neurosciences try to connect to a broader audience, what strategies do they adapt from literary practice? How do the ideas they ponder or promulgate inform their own style and voice on the printed page? What can we say about this kind of writing as a presence in our literary culture—possibly as a genre in its own right?

If classic spiritual or metaphysical descriptions of the mind could be challenged, in the public sphere, by new descriptions based on biochemistry, physiology, computation, or other studies and enterprises related to the brain and consciousness, what might the aesthetic and moral implications be of a shift like that?

What impact might we see upon the writing we produce, in the humanities, *about* the experience of reading and witnessing—in other words, what are the consequences and challenges for literary theory and for critical practice?

One other preliminary seems important: to recognize how certain key terms banging around in everyday parlance, and also in humanities discourse, are replete with clouds and contradictions. It is helpful to begin with definitions borrowed from *the Oxford English Dictionary* and various online authorities: *mind, brain, thinking, personality, identity, self, soul.* For some of these terms it's worth considering the etymology; for others, we can marvel at the variety of accepted meanings and at the discrepancies among them. Such an exercise is loosely Socratic as an opening subversion. For a next step, there are good reasons to look at short readings from around 1800, highlighting Coleridge's affirmation that the moment of imaginative insight is a moment of conjoining with "the Infinite I AM"; Marx and Engels and the proposition that consciousness and the self are economic and political constructs; Walter Pater's celebration of the "hard gemlike flame" of free and absolute imaginative intensity. However, if such preliminaries and backgrounds go on too long, students could lose track of what is really at stake; and because the conversation between the great brothers William and Henry James, with regard to consciousness and personal identity, provides such a foundational moment in our modern cultural history, and also a literary case study without parallel, it's worth going to them straight away, to explore and dramatize core issues in the class.

When I taught the course recently I worked with crucial chapters from William's *Psychology: The Briefer Course* ("The Stream of Consciousness," "The Self," "Imagination," and "Perception") and then took up Henry's *The Turn of the Screw*, "The Jolly Corner," and "The Beast in the Jungle." As an icebreaker for thinking about the questions I listed earlier, students were asked to write a short paper about one passage from any of Henry's narratives, discussing how that passage creates an impression or representation of the dynamics of inner life, the feel and fabric of human thought. Classroom discussions can center on questions and problems that vexed both brothers: the confinements and perils of subjectivity, the contingency of personal identity, the limits of sensory and cognitive engagement with the world. And because a literature course is also inherently an experience in possibilities of language and rhetorical strategy, a class can spend considerable time comparing voices and speculating on relationships among bodies of knowledge and interest, personal temperament, and the sound and play of the words on the page.

Because such a sojourn with William and Henry James is an adventure in nuance, it makes sense to follow with a change in tone, a dive into texts and controversies that students may see as more overtly dramatic and immediately accessible. Accordingly, the class can engage Marx and Engels on the "economic sources of consciousness" (including the assertions from *The German Ideology* that consciousness is a function of economic circumstance), consider the general condition of mind-science in 1846 when the book was first published, and speculate about reasons for the durability of these propositions, even amid a subsequent flood of psychological and neuroscientific research. Afterward, Stanton's "The Solitude of Self" provides a powerful rejoinder to Marxian propositions. To a roomful of male politicians, one aging woman describes the nature and limits of the human self in language that uncannily echoes moments in William James, affirming that consciousness is indeed contingent upon worldly training and economic imperative, yet also holding that the individual mind can be capable of depths and moral intricacies on a scale that *The German Ideology* does not allow. A conversation about Stanton can also move into speculations about the gendering of perspective and voice in the nineteenth century, and about male pundits and their penchant for all-encompassing generalizations. This poised, quiescent independence in these words of Stanton, after a life of struggle for American women, may cause some students to make comparisons to Virginia Woolf's *A Room of One's Own*, to moments of sovereignty and sequestered pride in Emily Dickinson, and to other modern women writers contributing to an alternative ethos, a different poetic, a feminist imagining of the private self.

It is easy to transition from this clash of temperaments into the *fin de siècle* combat that pitted mal-practitioners of "neurology" (again, nearly all of them male) against women as a gathering threat to the social and economic status quo. Because most students know little of what passed for therapy in Gilded Age sanitariums and in the upstairs suites of upper-class private homes, moments from Beard's *American Nervousness* can foster an ardent exchange about "rest cures," sensory and social deprivation in the name of recovery, an era of dangerous and debilitating remedies for "hysteria" in the pre-Freudian sense of the term, and a heyday of unregulated and liberally administered elixirs laced with alcohol and laudanum. Such a moment in American medical and cultural history can support an encounter with larger-scale questions transcending any defined era. Without dwelling overlong in the ugly domains of Lombroso, Nordau, eugenics, and pseudo-Darwinist classifications of consciousness on the basis of race, gender, and nationality, we can turn to the opening chapter of Frank Norris's *McTeague* as one instance of an exuberant American novel infatuated with mind-science formulations that turned out to be not only baseless but also destructive; and the conversation widened to recall and consider other works in the canon that owe some kind of imaginative allegiance to flat-earth, Ptolemaic, or other constructs long since discredited. When the conversation turns to Kate Chopin's classic novel *The Awakening* and Gilman's unsettling short story "The Yellow Wall-Paper," students will be ready to discuss these narratives as acts of resistance, affirmations of something close to Stanton's doctrine of inviolate solitude—against incursions of what passed, at that time, for psychological expertise.

Chopin's heroine Edna Pontellier eventually drowns herself in the Gulf of Mexico rather than be subjected to passive-aggressive invasions of her privacy by the men in her Grand Isle confinement; Charlotte Gilman's unnamed young wife rips the hideously cheery paper off the walls of her supposedly comfy and therapeutic prison, prostrating her husband with shock; and some students will value these gestures of intuitive or commonsense refusal amid an age of mind-science authoritarianism. Freud excerpts can come next, along with at least one of the twentieth century's consummately Freudian imaginative writers. Moments from *New Introductory Lectures on Psychoanalysis* (1933) and *An Outline of Psychoanalysis* (1940) can bump us into modern conventions about the unconscious and the organization of the self; but because Freudian theory has composted over the past forty years, the encounter with D. H. Lawrence's lectures can provide a surprise. Canonical now in his own right, Lawrence has lots of Freud-fueled subversive fun with gray eminences in the American canon—his rants about Franklin, Hawthorne, Melville, and Whitman are highlights on the tour—and in ventilating this way he enacts what it means to live in the presence of fresh and overpowering ideas about the human mind. On such an excursion with Lawrence, students can enjoy his witty irreverence about authors that some of them have read recently, and perhaps ceremoniously, in college survey courses or back in high school AP English. Moreover, though as a touchstone for literary criticism Freud can seem quaint now—as the first big-scale idea system to sweep through scholastic humanities departments in the twentieth century—this encounter with Freud and Lawrence can spark discussions about the possible risks of seduction by one body of thought about the mind. In one of our sessions, a student suggested a neat analogy, lifted from the same decade as Lawrence's *Studies*: the sequence from Charlie Chaplin's *Modern Times*, in which Charlie as the Tramp, having spent hours on an assembly line tightening sets of hex nuts with a pair of box wrenches, stumbles out onto the sidewalk during a short break, glassy-eyed, and ready to box-wrench anything that crosses his path.

Onward to present discontents: when excerpts from Daniel Dennett and Steven Pinker are taken together, they can foster more comparison of voices and discussion of tone—specifically the demeanor of cool and genial detachment favored by these and other world-class authorities on contemporary neuroscience and its implications, as they breeze through existential dilemmas that have vexed philosophers, theologians, and literary artists for centuries. For instance, here is a bit of Pinker's *NPR*-style bemusement about what he calls "the illusion of the unified self."

> The spooky part is that we have no reason to think that the baloney-generator in the patient's left hemisphere is behaving any differently from *ours* as *we* make sense of the inclinations emanating from the rest of *our* brains. The conscious mind—the self or soul—is a spin doctor, not the commander in chief. Sigmund Freud immodestly wrote that "humanity has in the course of time had to endure from the hands of science three great outrages upon its naïve self-love": the discovery that our world is not the center of the celestial spheres but rather a speck in a vast universe, discovery that we are not specifically created but instead descended from animals, and the discovery that often our conscious minds do not control how we act but merely tell us a story about our actions. He was right about the cumulative impact, but it was cognitive neuroscience rather than psychoanalysis that conclusively delivered the third blow. (45)

But if Dennett and Pinker seem incapable of angst about the philosophical and moral implications of possibilities in contemporary neuroscience, Paul Broks has plenty to share.

> My area of supposed expertise, neuropsychology, is the subject about which I feel the most profound ignorance. . . . When it comes to understanding the relationship of the brain and the conscious mind, my ignorance is deep and there is nowhere to turn.
>
> An ocean of incomprehension heaves beneath the textbook-confident surface of plain facts and technicalities that I present to my colleagues and patients. I have a clear picture of the material components of the brain and am prepared to ad lib at length about features of its functional architecture—the interlocking systems and subsystems of perception, memory, and action. But quite how our brains create that private sense of self-awareness we all float around in is a mystery. I have no idea how the trick is achieved. (91–92)

At several places in *Into the Silent Land*, Broks offers up confessions and insights that can chill the exuberance of people who believe that breakthroughs on the gravest matters related to consciousness will soon transform the retail business of literary scholarship and cultural studies.

> Debate is lively, sometimes strident, and with the neuroscientists shouting loudest of all above their noisy brain scanners, most do not notice

the fly buzzing frantically to escape the fly-bottle. They are engrossed. How *does* the mental arise from the material? How *can* the subjective experience be reconciled with that soggy mass occupying the skull? They are full of confidence, too. Most of them expect a solution. The chimera of consciousness rises like a vapour and entices them to believe that it really is just a matter of time before a way is found of accounting for the subjective, first-person phenomena in objective, third-person terms. Despite the prodigious amount of intellectual energy that has been driven into this enterprise in recent years, philosophical and scientific, it seems to me that the fly is still stuck in the bottle. (96–97)

Broks's account of a patient with a malignant brain tumor opens dangerously with a variation on Kafka's "The Metamorphosis" and closes in raw nightmare as the havoc wrought by the butterfly glioma destroys the victim's last grip on reality; in a later chapter, Broks offers a darkly comic interrogation of someone (who may or may not be himself) by stewards of a vague and sinister Academy, on a charge that embers of humanist hope continue to glow deep within this forensic scientist. There are also meditations, playful and somber, on Robert Louis Stevenson's "The Strange Case of Dr. Jekyll and Mr. Hyde," on Edwin Abbott's *Flatland*, on Robert Frost and a poetry of consciousness. All of this adds to an impression of an author willing to deploy everything he has learned, from the arts and wisdom of the West, to achieve some kind of leverage over recognitions from his own specialization, recognitions that threaten to swallow him whole. The chapters shift unpredictably, drastically, even desperately, in subject, voice, frame of reference. This is a narrative about struggling to keep an empowering citizenship in both of the cultures that C. P. Snow mapped out more than fifty years ago; and students who are troubled by *Into the Silent Land* may write with special eloquence about their own engagement with Broks.

In Damasio's *Looking for Spinoza*, a quest with geographic and historical dimesions, readers unsettled by Broks can find a salutary lightening of mood. The historical parallels that structure Damasio's *Spinoza* are elegant: on the loose at a neuroscience convention in Amsterdam, and feeling acutely his identity as an exile of sorts, Damasio, as an American of Portuguese and Jewish descent, and giving his career to a study (brain injury research) that is in itself an esoteric and isolating heresy, takes lonely excursions away from the meetings, in a quest to find traces of Baruch Spinoza (1632–77)—also Portuguese, also Jewish, and also an exile, a pursuer of studies and contemplations that got him in trouble with mainstream thinking and authority in his own time. If describing the quest this way implies that there is a subtext of self-congratulation in Damasio's account, nothing like that breaks through. Countering any such inference is Damasio's skill in modulating his own voice, sustaining an impression that the bygone philosopher and the modern man chasing him through these byways are both intellects and temperaments turned outward, toward the truth of the human condition, rather than inward, upon their own illusory importance.

I mentioned earlier my interest in an emergent metaphor in several of these contemporary texts: the labors of authorship as an analogue for a fully functioning human brain and personal identity. Dennett and Damasio both develop that analogy, and in their comments on the possibility interesting echoes can be noted, suggestions that certain recognitions, anomalies, and paradoxes have resonated down from one century into the next, perhaps without these latter-day, scientifically sophisticated investigators and theorists seeing the kinship. For example, a famous observation from William James in *Psychology*—"the provisional solution which we have reached must be the final word: the thoughts themselves are the thinkers"—seems embedded in Dennett's declaration near the end of *Consciousness Explained* that "we don't spin the tale, the tale spins us" (418), as well as a Damasio summation from his 1999 best seller *The Feeling of What Happens.*

> You know it is *you* seeing because the story depicts a character—you—doing the seeing. The first basis for the conscious *you* is a feeling which arises in the re-representation of the *nonconscious proto-self in the process of being modified* within an account which establishes the cause of the modification. The first trick behind consciousness is the creation of this account, and its first result is the feeling of knowing. (172)

The echoes are oblique and complex, and they do not support any *plus ça change* disjunction, allowing us to shrug off research in the neurosciences as bringing in nothing new. What can be contemplated and enjoyed here is the persistence not of an old idea but rather of classic imaginative strategies—reifications of irony, paradox, narrative structure, and the creative impulse and act, as ways of refreshing our understanding of human nature.

A final gathering of books on the syllabus is straight-out fiction: recent novels responding in various ways to realities and possibilities rooted in the contemporary neurosciences. Obviously the range available to us now is huge—from out-ahead dystopias in *The Matrix* mode (and devolving now into cliché), to original and meditative works anchored in our own moment, narratives about ordinary people, working scientists, eager students, grandstanding popularizers—all entangled, in one way or another, with existential problems about the brain, the mind, the body, and the self. The inclusion of a William Gibson novel can be a setup: though some students might be energized by a dive into familiar apocalyptic action, pleather trenchcoats, enigmatic robo-women, and weird techno-weaponry, others might recognize how quickly their own taste can ebb for this kind of thing, and send them looking for old-school literary qualities like character development and complex moral situations. So with a move from *All Tomorrow's Parties* to Richard Powers's *The Echo Maker*, a recent winner of the National Book Award for Fiction, the conversations can intensify and loosen up. It's hard to think of a contemporary novel that considers so broadly and thoroughly the present configuration of neuroscientific research and its implications for the human condition. Powers begins with a mysterious one-truck accident at night in the broad valley of the Platte River in central Nebraska, in a season when thousands of sandhill cranes are moving through these wetlands on a migration that in itself signifies a compelling mystery about knowing and finding one's way, about the mind, individual, and collective identity, and understanding or intuiting one's place in the scheme of things. The accident triggers a case of Capgras delusion, a rare neurological disorder in which the victim cannot be dissuaded that someone close to him is an impostor, a changeling—in this case his own loving and caring sister. Expanding from there in widening circles, the novel is informed, meditative, and elegiac in ways that I'd prefer not to boil down here; Powers does his homework with completeness and ferocity, and on any short list of the best fiction engaging neuroscience in the early twenty-first century, *The Echo Maker* would probably rank near the top.

With a very different feel, Dan Lloyd's *Radiant Cool* is a hectic, giddy narrative suggesting the advantages and risks of being in the thick of this research and cultural ferment: it tries exuberantly to rope in the transient latest (2004) in neuroscience theory; containing a comic whodunit that runs for quite a distance on e-mails and online archives, it breaks out in cadenzas of graphic-novel and *animé* spinoff—even a Great Conspiracy by a kind of neuro-cyberpunk Doctor Evil.

> We will fold our agents into every interchange, and none will be able to organize against us." I imagined ranks of glowing pods in a shadowy underground warehouse, each one housing duplicates, triplicates, multiplicates of plain folks. Suddenly all their eyes spring open. That would be the film version, of course. The bunker Porfiry had in mind was just a room full of computers, the wake-up call a keystroke. The invasion, a

flood of words, words, words. But marching out, swarming the globe, *spam consciousness*. (176)

There are plenty of bells and whistles, including an avatar of Dan Lloyd himself who eventually turns up to shed light on the foreground mystery and surmise at length about the organization of the human mind. *Radiant Cool* is a mix of goofy exuberance and paranoia, a headlong confusion of excitement, humor, and fear about what might await us all, fifteen minutes down the road. Students will recognize that tone, and they will want to discuss it as a demeanor, or perhaps as a personal and cultural strategy of evasion, with regard to the formidable subject under scrutiny here, the nature and consequentiality of the mind. But because the novel runs long, and because intricacies of the plot seem less important than the speculation-cadenzas that Lloyd loads into his narrative, it might be well for a class to read around in it rather than rumble from the first chapter to the last.

As everyone knows, headlines from the neurosciences fuel no end of pop-culture action, and sensational possibilities are showing up constantly in big-budget mass entertainments. A conversation about that material can suggest stronger connections between the past and the present, the esoteric and the everyday: fantasies pumped out by Hollywood, the game-building industry, the online miasma. Material suggested by my students included the quirky, uneven *Caprica*, a SyFy Channel series that has already made its contributions to cyberspace folklore, including a dubious gone-viral "fact" that the storage capacity of a human brain is only about 300 megabytes of memory, and also that as data about everyone piles up on the servers, personal identity, every nuance of it, will soon be reconstructable on a drive. Online chatter about the show buzzes that this "300 megabytes" phrase was silently amended to "300 terabytes" in a later printing of the Season One DVD, perhaps as appeasement for those of us who want the synaptic apparatus in our own skulls to be more dignified than the equivalent of one disk. As I write, *Caprica* may be spinning out of control: for instance, it features a cult that has embraced the dream of the downloadable self as a shortcut to personal immortality—resurrection now!—thereby eliminating the bother and expense of theology and faith. Not such a tractable subject for a show with clanking robots and whizzing spaceships. We also bumped into Joss Whedon's *Dollhouse* (in which comely, dangerous young folk get their brains regularly reprogrammed, the mind rebooted every week by another Big Conspiracy); and at this writing *The Sarah Connor Chronicles* is still going strong as the latest iteration of the Terminator franchise. The summer of 2010 brought us Christopher Nolan's ponderous

Inception, about mind-control and levels of consciousness blueprinted like an IKEA layout—and a spate of e-mails from former students, insisting that I sit through it—in sort, plenty to choose from, and so much more in the pipeline.

If this nearly final phase for such a course sounds like a plunge into the imitative anarchy of the pop, it doesn't have to be pandering. For any study with historical arcs to it, one collateral impression to countenance, and resist, is that everything was smooth and more or less orderly until this morning, that ages and eras and cultural fashions and vogues have swept through the culture one at a time, and that these street parades never fall apart until the very end, when we ourselves arrive at the curb to bear witness. To the dialogue with ourselves about where we are and where we might be heading, a basic contribution of the humanities has been richer and truer reckonings between the past and the present, a heightened sense of how our culture has found its way, or blundered into these circumstances, what memorable reactions and negotiations have been triggered in the recent past, and where and how these conversations have moved on. In text and art, cultural responses to change and crisis are a classic humanities subject; and if there are assumptions out there that new research and paradigms about creativity and consciousness will inevitably knock the whole enterprise into permanent obsolescence, that notion strikes me as absurdly wrong. If we mean to keep faith with core intentions and aspirations that brought this inquiry into shape—a deeper understanding of the present, achieved with fresh attention to a scientific and cultural past—then we have a journey ahead, and we can enjoy the ride.

WORKS CITED

Abbott, Edwin A. 1925. *Flatland: A Romance of Many Dimensions.* Boston: Little, Brown.

Beard, George. 1881. *American Nervousness.* New York: G. P. Putnam's Sons.

Broks, Paul. 2003. *Into the Silent Land: Travels in Neuropsychology.* New York: Grove Press.

Chopin, Kate. 1899. *The Awakening.* Chicago: Herbert S. Stone.

Damasio, Antonio. 1999. *The Feeling of What Happens: Body and Emotion in the Making of Consciousness.* New York: Harcourt, Brace.

Damasio, Antonio. 2003. *Looking for Spinoza: Joy, Sorrow, and the Feeling Brain.* New York: Harcourt, Brace.

Dennett, Daniel. 1991. *Consciousness Explained.* Boston: Little, Brown.

Donald, Merlin. 2001. *A Mind So Rare: The Evolution of Human Consciousness.* New York: W. W. Norton.

Faulks, Sebastian. 2005. *Human Traces.* New York: Random House.

Freud, Sigmund. 1955. *An Outline of Psychoanalysis*. Edited by Clara Thompson and James Samuel Van Teslaar. New York: Modern Library.

Freud, Sigmund. 1966. *Introductory Lectures on Psychoanalysis*. New York: W. W. Norton.

Gazzaniga, Michael S. 2008. *Human: The Science behind What Makes Your Brain Unique*. New York: HarperCollins.

Gibson, William. 1984. *Neuromancer*. New York: Ace.

Gibson, William. 1999. *All Tomorrow's Parties*. New York: Ace.

Gilman, Charlotte Perkins. 1899. *The Yellow Wallpaper*. Boston: Small and Maynard.

Gladwell, Malcolm. 2005. *Blink: The Power of Thinking without Thinking*. Boston: Little, Brown.

James, Henry. 1998. *The Turn of the Screw and Other Short Stories*. New York: Oxford World Classics.

James, William. 1986. *Psychology: The Briefer Course*. Notre Dame: University of Notre Dame Press.

Lawrence, David Henry. 1990. *Studies in Classic American Literature*. New York: Penguin Editions.

Lehrer, Jonah. 2007. *Proust Was a Neuroscientist*. New York: Houghton, Mifflin.

Lloyd, Dan. 2004. *Radiant Cool*. Cambridge: MIT University Press.

Marx, Karl, and Friedrich Engels. 1965. *The German Ideology*. London: Lawrence and Wishart.

Piaget, Jean, and Barbel Inhelder. 1966. *The Psychology of the Child*. New York: Basic Books.

Pinker, Steven. 2002. *The Blank Slate: The Modern Denial of Human Nature*. New York: Penguin Putnam Books.

Powers, Richard. 2006. *The Echo Maker*. New York: Farrar Straus and Giroux.

Stanton, Elizabeth Cady. 2006. "The Solitude of Self." Washington, DC: National Park Service/Eastern National Publishing.

Whitehead, Alfred North. 1925. *Science and the Modern World*. New York: Free Press.

Woolf, Virginia. 1925. *A Room of One's Own*. New York: Harcourt, Brace.

Afterword

Twisting the Neurohelix

Joseph Dumit

This book indexes a new neuroscientific turn in what we probably should think of as the neurohelix. If the first twist was neuroscience, the second twist the critical analyses of neuroscience by science and technology studies, the third twist the turning to humanities topics by neuroscientists, and the fourth twist the appropriation of neuroscience results into humanities and social sciences relatively uncritically, then this fifth twist is the mutual engagement of these groups, trying to sort out what we actually have learned from all of these studies of brains as if they were persons, and studies of persons as if they were brains. If we want to pursue the analogy by which neuroscience becomes another molecular biology, promising to unravel the secrets of human nature, then we might think of this moment as akin to when we sequenced the human genome, only to realize that genes provide precious little guidance, requiring quite a bit of developmental help in order to make proteins, whose functions are not yet mapped, and we have yet to get to epigenetics where "environment interacts with identical genotypes to produce a variety of phenotypes." In other words, despite the wonderful correlations of brains with behaviors and types of persons, they are still just that, correlations, all too loose, and as Whitehouse's chapter on Alzheimer's illustrates, often mystifying their correlates more than clarifying them.

Neuroscience Requires Translation

Similar to the project of "critical neuroscience," the nascent critical encounter between the neurosciences and the humanities that this book stages can help if we take the time to let each author engage all of the other ones. This means taking up the challenge of neuroscience as a "translational discipline," as the editors have so nicely termed it. As a group these chapters call for carefully evaluating every word of every sentence of a neuroscientific claim. Beaulieu and Fitzpatrick remind us that the tendency to take a popular science news account of an experiment, or even the abstract, at face value is to turn ourselves into more cognitive paparazzi, repeating neuroscience findings if they were "accessible and readily interpretable by nonexperts." Almost every chapter here reminds us that it is important to look critically at the original articles for many reasons: the first is that the results may not be what we think the abstract, science reporting, or the conclusion claim it to be. Second, the scope of application or generalization of the task, the aspect of human nature, or the strength of the results may be quite different from our uses for them. And third, we can push back and ask for better experiments. This is critical. This holds for sentences as well as images.

Neuroscience translation is complicated above all by the technical use of words in experiments, *everyday words* like *men, women, honest, experience*, that are proxies for specific selection processes and highly designed protocols, words that nonetheless blur into their everyday connotations; *controversial words* like *brain activation*, when measured changes above a threshold is meant, or *executive function* as if it can be taken for granted that there is well-defined relationship between a brain region's measured change and *execution* as a stable term; and *individualizing language* like "honest subjects showed no signs of engaging additional control processes," implying something about *all* honest subjects when only group averages are meant and there is likely to be substantial overlap among the two groups of subjects being compared.

One way we can understand how this happens is by analyzing how neuroscience journal articles, and neuroimaging articles in particular, begin from or build upon social categories of personhood like adolescence or criminality (or female or smart or gay or Alzheimer's, etc.). In so doing, they often show a strong tension between scatterplots, images, and claims. Put concisely, individual data show large variation within a group that overlaps, often entirely in range, with the large variation within the contrasting group. In other words, scatterplots make it clear that, for instance, given a score in the brain measure of interest, there is no way to claim which group a subject is in. At the same time, there are often statistical trends in the data that suggest there may be something interesting to investigate further. This type of work is best described as hypothesis generating.

Yet when images or conclusions are made, they almost always reduce the two categories to the trend, rendering them absolutely opposed to each other visually. I have argued in *Picturing Personhood* (2003) that this type of "extreme image" produces a polemical image that makes claims often at odds not only with the data but also with many of the textual claims made in the articles. One terrible problem is that not enough papers provide scatterplots that make clear the overlap (or potential lack of it), or if they do, those scatterplots are of a composite measure, sometimes conflating a behavioral measurement like reaction time with a brain measurement, as if the two are naturally coupled. Neuroscience needs more comprehensive data presentations of individual subjects. For those of us who want to use neuroscience and translate its terms into our own disciplines, we must demand this data in order to understand how it can or can't be generalized, before accepting the claims.

Neuroscience Is Not Yet a Boundary Object

Some of the chapters in this volume are experimenting with importing neuroscience findings directly into their home disciplines such as history, literacy studies, and ethics. The primary advantage of this is that in doing so, they expose the hidden biologism inside of many humanistic and social science projects: the assumption that we can talk about people, humans, as if they are the same. As Elizabeth Wilson, Elizabeth Grosz, and other feminist analysts have pointed out, dividing the world into genders, sexes, races, ethnicities, or sexualities doesn't help much, since one still imagines two kinds of knowledge, one about a social world and the other about a biological one. Engaging with biology, including neuroscience, and even collaborating with biologists and neuroscientists, are crucial. But the next step, as suggested by Kismet Bell, Gotman, and Murison, is to remain a humanist and interrogate the history, models, and metaphors being used, in the formulation of the problems within neuroscience, in the experiments, and in the result representations. And Beaulieu reminds us that the expertise of the humanities is in close reading, and we must practice this continually.

For example, the Hendrix and May chapter is fascinating for its distinguishing between two very different mystics with different mystical experiences, gained through different practices at different times in different cultural settings and under different theological regimes. They provide excellent descriptions of the direct connections between practices (like sleep deprivation and meditation) with visions. I'm not sure what is gained by saying that each of their experiences was "likely mediated by" a brain structure, however, because we cannot know, and because they choose to take contemporary neuroscience as ahistorical, leading to the following sentence.

> Historians have made great strides in gaining an increasingly sophisti-
> cated understanding of peoples of the past who lived very different lives
> from the researcher and explained their own experiences through radi-
> cally different discursive formations—yet were *dependent upon the same*
> *neurological structures* for forming thoughts and approaching the world
> as we are today. (emphasis added)

What is gained by the phrase "the same neurological structures"? It seems as if the opposite claim could be equally true: that different times, cultures, belief systems, and practices lead to different neurological struc-tures. The word *structure* only appears twice in that paragraph referring to "standardized neurological structures" as a presupposition of neurohis-tory, and then later once more, referring to the "power structures" in soci-ety that are treated as nonuniversal. There is not time in an epilogue to do more than pose questions for neurohistory: What is a structure? Do structures change over time and place? Are they plastic? What about the structures of science? I'm quite serious about these questions. We need to consider scales of analysis. A neurohistoriography must begin with an enquiry into the history of its own universals, both those borrowed from history and from neuroscience. This practice includes a way of accounting for its own sense that neuroscience has finished thinking about neurologi-cal structures, when most neuroscientists do not. How is a neurohistory to take into account the history of neuroscience, such as Kismet Bell or Murison?

Gorzelsky provides one counterpoint in her careful study of the effects of meditation on revising one's view of oneself. But despite the marshaling of much brain research for how meditation might change brain biology, a similar problem of universals arises in her text: "For instance, acquiring the literate practices described above [meditation] probably *effects one set of neurological changes* with significant social and political implications, while acquiring those of a racial supremacist group probably *effects another set*, with equally significant implications" (emphasis added). If we already know that meditative literate practices have one kind of social and political impli-cation, and racial supremacy practices have another, my first response is that we should spend our time teaching meditation and not racial suprem-acy and not waste money on figuring out what neurological changes are correlated with each. The claim that there must be a difference is either tautological—that every practice causes a different brain change—or it is unfounded—since nowhere in the article is any evidence at all discussed regarding racial supremacy. Would it be beyond belief, for instance, if it turned out that chants and extended concentration on one supreme race effected the same neurological changes as compassionate meditation? I wager that it would not change Gorzelsky's position on the value of the latter over the former. I could certainly see some neuroscientists insisting that they just hadn't looked in the right places then, and the changes must be located at a different level of resolution or among as yet undiscovered neurological structures. But my challenge to Gorzelsky is for her to apply her literacy studies to neuroscience: What are the social and political implications of the literate practices of neuroscience, for practitioners and those who translate their work? If meditation and racial supremacy literate practices are our ethical beacons, can we evaluate neuroscience, including its meditative, racial, medical, and military sides? And to take her claim above seriously, would it make a difference if it turned out that neuroscientists had one or the other set of neurological changes? If this question seems silly, please consider why the initial claim did not. I think the fate of this neuroscientific turn turns on our ability to consider what happens when neuroscience results challenge our stereotypes and status quo.

Neuroscience Is Changing the World

Other chapters in this volume focus on the performativity of neuroscience, on the ways in which theories of neuroscience are enacted in practices that produce the worlds that the theories talk about. This may sound like social constructivism, but as STS theorists like John Law and Annemarie Mol have explained, these are eminently material practices. Essays by Whitehouse and by Rosenquist and Rothschild trace the paths through which neuroscientific theories are taken up in clinical medicine and in economics to reify certain positions over others. Whitehouse recounts the material construction of an Alzheimer's industry built in no small part on inflated neuroscientific claims. He suggests that the industry, besides being growth and hype oriented has dangerous consequences not only for patients, families, and protopatients who live in fear of some "thing" but for clinical research in general, where resources are directed toward areas that promise the most rather than those that can make lives better. Neuroscience has, in other words, brought into being a set of institutions that are unable to come to grips with its major finding: "As a result of clinical and pathological variability, there are no clear biomarkers that qualitatively discriminate

Alzheimer's disease from normal aging." Rather than this being a sign that society might better pursue other paths, the neuro-logic is twofold: (1) if we haven't found a biomarker, we surely will with more money and better instruments; and (2) pretend as if the preliminary findings we have are enough to make an industry out of risk "diagnostics" and "preventive" treatments. It is here that Whitehouse suggests a different approach from within neuroscience: "Perhaps we need to consider reevaluating the way we think and talk about some brain diseases. This should be a fundamental issue for a broader neuroethics." I am fascinated by this neuroethics that begins with a mode of reflection on our own fascination with certain solutions.

Rosenquist and Rothschild see a similar neurorealism driving the rapidly coalescing field of neuroeconomics. Connecting neuroscience research to economic problems provides what they feel is an illusion of objectivity to what are essentially values-based claims. Being an illusion does not preclude it from successfully buttressing an argument in the eyes of readers, however. "The risk posed by normative neuroeconomics," they fear, "is that it can be used to obscure— indeed, to obfuscate—the implicit value judgments involved in reaching public policy conclusions" (14). Hidden assumptions gain the shiny veneer of neuroscience, and one group's values potentially gain the status of fact. They are quite adamant about this problem: "It is precisely this sort of appeal to neuroscientific evidence as a substitute for more direct value judgments that makes us nervous" (21). Hidden assumptions gain the veneer of science, and one group's values potentially gain the status of fact. They are quite adamant about this problem: "It is, in our opinion, a mistake to read existing neuroeconomic evidence as scientifically verifying any particular descriptive behavioral economic models."

In light of the above critiques, we might reconsider one of the conclusions of Racine and Zimmerman, as we did with Gorzelsky, asking where the prescriptive force of their claims arise from.

> For example, if highly emotional ethical decisions were *handled by different neuronal subsystems* than less emotionally charged decisions . . . then practical ethics could benefit from including further consideration of this aspect of moral decision making. The handling of case discussions by clinical ethicists could *take explicitly into account the emotional state* of patients, family members, and providers when initiating discussions. (Emphasis added.)

On one hand, it seems obvious to me that clinical ethicists *should* take the emotional state of those they are talking to into account when discussing cases with them. I find it hard to imagine that ethicists could be trained otherwise. But if it is the case that they are not, and there is a dispute among ethicists, then the "fact" that decisions are handled by "different neuronal subsystems" does not add to the dispute (see Fitzpatrick). If they were not handled by different subsystems (Meaning what? That highly emotional decisions were handled the same way by the brain?), then the case for taking into account the emotional state of patients *remains unchanged.* Here I learn from another part of Racine and Zimmerman's chapter that "the process of 'inquiry' must include the agent's subjective experience and the input of other disciplines (e.g., anthropology, sociology) and research approaches (e.g., qualitative research) that provide broader and grounded perspectives, especially on the social context of decision making." Surely the evidence of the *effectiveness* of discussions that take emotional state into account, versus those discussions that do not, is what matters here, not whether the same or different neuronal subsystems are involved.

I belabor these arguments because so much is at stake. Neuroscience is increasingly deployed in courtrooms, for instance, as in 2004 when the American Psychiatric Association, the American Medical Association, and other large groups of scientists weighed in on a Supreme Court case involving the question of whether or not a seventeen-year-old was mature enough to be subject to the death penalty. Hundreds of passages like those I examine above were deployed, claiming that neuroscience studies showed conclusively that seventeen-year-olds had immature brains and therefore could not be fully culpable. However, an equally compelling set of neuroscience studies had been deployed over the previous thirty years to show conclusively that dangerousness and riskiness were inherent in some people, including, for instance, that brain scans of murderers look completely unlike brain scans of normals. My point is that the marshaling of neuroscience to make social claims only appears to be a good idea, or even to make sense, when one thinks that the claims are inherently good or obvious. Turning neuroscience claims around to ask whether the statements could survive a different brain result is just one way to understand the allure of neuroscience in our own justifications.

Brains Should Be Boundary Objects

A boundary object (as defined by Griesemer and Star, e.g., a map) allows different communities of practice to coordinate their activities. If the neurohelix is to keep turning, then claims about brains need to be examined word for word in order to coordinate them across disciplines. Neuroscience studies of social categories of difference do not strengthen or give bioevidence of the categories. If anything, they almost all weaken those categories by the sheer fact that despite looking and looking, they are unable to produce biomarkers. As in the case of Alzheimer's, this is surprising! Whether we take this to mean that we should keep looking, or that our initial selection procedure or diagnosis needs revising, or that we should put our resources elsewhere, we should never repress negative results, nor hide the lack of power in the positive ones. Social categories (the meanings of most words in most statements about the brain) have almost nothing in common with the technical terms (proxies) used in experiments. This is by design, but to what end? The cost is an ever-increasing vagueness in results, creating mystifying appropriations across disciplines rather than creating the grounds for rigorous discussions of the limits of the kinds of questions we are currently able to pose. Treating brains as boundary objects means welcoming discussions that might slow down the work being done on all sides. For instance, Hendrix and May's nuanced discussion of mystical experiences might allow us to recognize the limits of current experiments, and design ones that not only speak directly to the history of experiential terms but also interrogate the meaning of *brain mediation* itself.

As the Decade of the Brain becomes a distant memory, too much neuroscience still reads defensively, satisfied to claim that the brain matters or reaching beyond the data toward a future in which loose correlations really were biomarkers. I think this is completely wrong. Of course the brain matters. What we really need neuroscience to do is be willing to explore the limits of the brain in understanding human experience as well. At this shared intersection humanists and neuroscientists are coexplorers and codiagnosers of our human condition.

Contributors

ANNE BEAULIEU works at the Groningen Energy and Sustainability Programme of the University of Groningen, The Netherlands. Before that, she was Senior Research Fellow at the Virtual Knowledge Studio for the Humanities and Social Sciences (VKS) of the Royal Academy of Arts and Sciences in Amsterdam. She has studied the development and consequences of imaging and databasing technologies for biomedical knowledge, including an ethnographic study of brain imaging in neuroscience in the course of her PhD, entitled "The Space inside the Skull: Digital Representations, Brain Mapping and Cognitive Neuroscience in the Decade of the Brain." Her research follows two main lines: the study of the use of databases and networks in knowledge creation, and the development of new ethnographic approaches to cultural and social phenomena in mediated settings. Together with Sarah de Rijcke (Leiden University), she recently pursued an ethnographic study of knowledge production around databases of images on the web entitled *Network Realism*.

JAMESON KISMET BELL is an assistant professor in the Division of Humanities and Social Sciences at Doğuş University in Istanbul, Turkey. He received MA and PhD degrees in German Studies from the Pennsylvania State University. He has taught special education in mental hospital and school settings, grades 2–12. He was awarded a Fulbright Fellowship to conduct research on the history of the brain at the Swiss Federal Institute of Technology (ETH) in Zurich, Switzerland. His dissertation explores the performativity of knowledge in early sixteenth-century medical and surgical approaches to the brain in a German context. His recent publications are both literary and academic.

SARAH BIRGE is a Lecturer in the Program for Communication Excellence at Rice University and a doctoral candidate in English at the Pennsylvania State University. Her work has appeared in *Disability Studies Quarterly* and *Journal of Medical Humanities*, and her research focuses on intersections of science, medicine, and disability in narrative. Her dissertation, "Surviving the Narrative Self: Cognitive Disability in Contemporary Fiction," examines ways in which cognitively disabled narrators construct and perform selfhood.

JOSEPH DUMIT is Director of Science and Technology Studies and Professor of Anthropology at the University of California Davis. He is the author of *Picturing Personhood: Brain Scans and Biomedical Identity* (Princeton University Press, 2004). Dumit has also coedited numerous collections, including *Cyborgs & Citadels: Anthropological Interventions in Emerging Sciences and Technologies*, *Cyborg Babies: From Techno-Sex to Techno-Tots*, and *Biomedicine as Culture*. He was associate editor of *Culture, Medicine & Psychiatry* for ten years. He has just finished a book on pharmaceutical marketing and clinical trials called *Drugs for Life: How Pharmaceutical Companies Define Our Health* (Duke, forthcoming). He has begun work on a new project on the history of flow charts, cognitive science, and psychotic computers.

SUSAN M. FITZPATRICK earned her PhD in Biochemistry and Neurology from Cornell University Medical College (1984). After initially undertaking NMR spectroscopic studies of brain metabolism in the Department of Molecular Biochemistry and Biophysics at Yale University, Fitzpatrick shifted her career to nonprofit administration. She is currently Vice President of the James S. McDonnell Foundation and also serves as an adjunct associate professor of Neurobiology and Anatomy at Washington University School of Medicine (St. Louis), teaching neuroscience. She comments and writes on advances in cognitive rehabilitation, on the role of private philanthropy in advancing scientific knowledge, and on issues at the intersection of scientific research and public understanding of science.

GWEN GORZELSKY is associate professor and Director of Composition at Wayne State University. She has published articles in *CCC*, *Reflections*, *JAC*, *JAEPL*, and other venues. Her 2005 book *The Language of Experience: Literate Practices and Social Change* (University of Pittsburgh Press) examines the relationship between literacy and change, both personal and social. She contributes to program and university-wide initiatives in community-based education and service learning. She is writing a book that combines qualitative research with historically situated textual analysis to consider how the literate practices of Zen teacher Thich Nhat Hanh work to promote neuroplasticity.

KÉLINA GOTMAN is a Lecturer in Theatre and Performance Studies in the Department of English at King's College London, and convenor of the MA in Theatre and Performance Studies. She holds a PhD from Columbia University and a BA from Brown University, and has taught cultural and critical theory, drama, and writing at Bard, The New School, Columbia, and KCL. She was Audrey and William H. Helfand Fellow in the Medical Humanities at the New York Academy of Medicine (2008–9), and an invited speaker at the first international Epilepsy, Brain and Mind Congress (Prague) and the History of Psychiatry Research Seminar at the Cornell-Weill Medical Centre (New York). She writes regularly on dance, philosophy, and the arts, and is completing a book on dance manias in nineteenth-century medical literature, looking at the discourse surrounding involuntary movements, epidemics, and revolution in early neurology and ethnographic literature (Choreomania). Her translation of Félix Guattari's *The Anti-Oedipus Papers* was published by Semiotext(e)/MIT Press (2006).

SCOTT E. HENDRIX holds a PhD in history from the University of Tennessee and is currently an Assistant Professor of History at Carroll University in Waukesha, Wisconsin. He has written on historical and cultural approaches to knowledge, with his most recent work being *How Albert the Great's Speculum astronomiae Was Interpreted and Used by Four Centuries of Readers: A Study in Late Medieval Medicine, Astronomy and Astrology* (Lewiston: The Edwin Mellen Press, 2010). More recently he has focused on the study of mysticism in collaboration with Christopher J. May, with ongoing research on a range of historical figures from Catherine of Sienna to Alisteir Crowley.

JUDY ILLES is Professor of Neurology and Canada Research Chair in Neuroethics at the University of British Columbia. She is Director of the National Core for Neuroethics at UBC and also holds an appointment as Adjunct Professor in the School of Population and Public Health at UBC and as a Faculty Affiliate of Department of Computer Science and Engineering at the University of Washington. Dr. Illes is a member of the Dana Alliance for Brain Initiatives, co-Founder and Governing Board member of the International Neuroethics Society, and past Chair of Women in World Neuroscience for IBRO. Dr. Illes's research focuses on the ethical, legal, social, and policy challenges at the intersection of neuroscience and biomedical ethics, including neuroimaging, commercialization of cognitive neuroscience, and the engagement of key stakeholders on a global scale. Dr. Illes has written numerous books, edited volumes and articles.

Her latest books, *The Oxford Handbook of Neuroethics* (J. Illes and B. J. Saha-kian, eds.), and *Addiction Neuroethics* (A. Carter, W. Hall, and J. Illes, eds., Elsevier Press), were both published in 2011.

JENELL M. JOHNSON is an assistant professor of Communication Arts and the associate director of Disability Studies at the University of Wisconsin–Madison, where she teaches courses on the rhetoric of science and disability studies. She is currently working on a book manuscript entitled *Medical Marvel: Lobotomy in American Medicine and Culture*, which explores how representations of lobotomy in American popular culture shaped the development and decline of one of the most infamous procedures in the history of medicine. Her work has appeared in *Rhetoric Society Quarterly*, *JAC*, *Medicine Studies*, and *Journal of Literary and Cultural Disability Studies*.

MELISSA M. LITTLEFIELD is an assistant professor in the Department of English and the Department of Kinesiology and Community Health at the University of Illinois, Urbana-Champaign. She is also an affiliate of the Beckman Institute for Advanced Science and Technology. She is the author of *The Lying Brain: Lie Detection in Science and Science Fiction* (University of Michigan Press, 2011). Her work has appeared in *Science, Technology & Human Values*, *American Literary History*, *Neurology and Modernity: A Cultural History of Nervous Systems, 1800–1950*, and *The Routledge Companion to Literature and Science*. She is currently the primary investigator for an fMRI study hosted by Aarhus University.

CHRISTOPHER J. MAY is an assistant professor of Psychology at Carroll University, where he teaches neuroscience courses and conducts research on the psychophysiological effects of meditation. His most recent work on meditation, "Short-Term Training in Loving-Kindness Meditation Produces a State, but not a Trait, Alteration of Attention," has been published by the journal *Mindfulness*. In addition, Dr. May develops neural network models to investigate the cortical underpinnings of meditation-induced cognitive changes.

BRUCE MICHELSON is Professor of English and Director of the Campus Honors Program at the University of Illinois. His books include *Printer's Devil: Mark Twain and the American Publishing Revolution* (2006), *Literary Wit* (2000), *Mark Twain on the Loose* (1995), *Wilbur's Poetry* (1991), as well as three editions of the *Instructor's Guide for The Norton Anthology of American*

Literature. Featured in episodes of the *American Passages* video series, he has been a Fulbright Scholar, a USIA-sponsored lecturer in India for the hundredth anniversary of Twain's *Following the Equator*, and a recent winner of the University of Illinois Distinguished Teacher/Scholar Award. He is President Elect of the American Humor Studies Association and Past President of the Mark Twain Circle of America.

JUSTINE S. MURISON is an assistant professor of English at the University of Illinois at Urbana-Champaign. Her book *The Politics of Anxiety in Nineteenth-Century American Literature* was published by Cambridge University Press in 2011. It explores the political, literary, and religious quandaries posed by studies of the nervous system in the nineteenth century. Her work has also appeared in *Early American Literature* and *Arizona Quarterly.*

ERIC RACINE, PhD, is the Director of the Neuroethics Research Unit at the IRCM (Institut de recherches cliniques de Montréal) and holds appointments at the University of Montreal (Medicine and Bioethics) and McGill University (Neurology & Neurosurgery, Medicine, and Biomedical Ethics). He is a New Investigator of the Canadian Institutes of Health Research and a principal investigator on several projects examining ethical and social issues in research and health care sparked by rapid advances in neuroscience and neurotechnology. He is the author of *Pragmatic Neuroethics* published by MIT Press and of several peer-reviewed papers and chapters on different issues in neuroethics.

JAMES NIELS ROSENQUIST is an instructor at Harvard Medical School and a research fellow at Harvard University's Institute for Quantitative Social Science. He holds both MD and PhD degrees from the University of Pennsylvania. A practicing psychiatrist, Dr. Rosenquist is a formally trained economist whose research focuses broadly upon the biological and environmental influences on economic behavior.

CASEY ROTHSCHILD is an Assistant Professor of Economics at Wellesley College. He holds a PhD in economics from MIT, and his research focuses on a variety of topics, including public finance and applied and evolutionary game theory.

PETER J. WHITEHOUSE is Professor of Neurology at Case Western Reserve University and Attending Physician at University Hospitals Case

Medical Center. He also holds or has held appointments in psychiatry, neuroscience, psychology, nursing, organizational behavior, cognitive science, bioethics, and history. He is developing an intergenerational health and wellness practice based on an integrative narrative evolutionary health model. His long-term interest is developing innovative clinical and learning environments to promote individual and collective health and wisdom. He is the author of *The Myth of Alzheimer's* (St. Martin's Press, January 2008; www.themythofalzheimers.com). He is a founder with his wife, Cathy, of The Intergenerational School—an innovative, successful, urban public school (www.tisonline.org).

EMMA ZIMMERMAN completed her bachelor's and master's degrees in bioengineering at Arizona State University. There, she worked with The Neural Engineering and Informatics Lab to develop a noninvasive diagnostic for coronary artery disease. In an effort to think more broadly about how these biotechnologies interacted with society she became interested in biomedical ethics. She joined the Neuroethics Unit in Fall 2009 where she is pursuing her PhD at the Graduate Program in Neurological Sciences at McGill University. Her research interests include the ethical and responsible transfer of knowledge in the area of social neuroscience.

Index

Abbot, Edwin, 227
affective neuroscience, 7, 137
Alzheimer's disease: biological under-
    standing of, 209–11; as a disease of
    the mind, 211; literature and, 98;
    myth of, 209; the neuroscientific
    turn and, 199–213, 237–38; other
    dementias and 204; social construc-
    tion and, 205, 234
American Association for the Advance-
    ment of Science, 193
animal magnetism, 31–33, 35, 39, 41,
    45n2. *See also* mesmerism
anthropology, 9, 20n5
anxiety disorders, 125, 131
Appelbaum, Laurence, 20n2
Arendt, Hannah, 74–75
artificial intelligence (AI), 219, 221
Atwood, Margaret, 21n14
Augustyn, Prisca, 20n5
Austin, James H., 112, 115, 117n8,
    124–25, 131n1
autism, 21n14, 94
Awakening, The (Chopin), 220, 222,
    224–25

Badiou, Alain, 77
Baudrillard, Jean, 221
Beard, George Miller, 30, 36–39, 42,
    44, 46, 221, 224
Beaulieu, Anne, 9, 18, 22n23, 50–51,
    120, 132n4, 152–59, 234–35
Beck, Ulrich, 11

behavioral economics. *See* economics,
    neuroeconomics
Benjamin, Walter, 84
Bergson, Henri, 77, 83
Bernard of Clairvaux, 107–13, 116
biocultural: defined, 122, literacy
    acquisition as, 123–25, meditation
    as, 124–27
bioethics, 135–36, 140, 143–44, 200–
    201, 203–4, 212. *See also* neuroethics
biological reductionism. *See* reduction-
    ism
biologism. *See* reductionism
biomarkers, 240; of Alzheimer's disease,
    210, 237, 240
biomedicalization, 43
BIOS Centre, 2
black box (Latour), 8–9, 12–13
BOLD fMRI (blood oxygenation level
    dependent functional magnetic
    resonance imaging). *See under* func-
    tional magnetic resonance imaging
boundary objects, 239–40
boundary work, 18
Bourdieu, Pierre, 114
brain event: defined, 50; of 1519
    Strassburg, 52–58; performative
    theory of, 58–61
brainhood (Vidal), 10, 22n14, 90–91,
    95, 99, 200. *See also* personhood;
    selfhood
brain scans. *See* imaging technologies
Brave New World (Huxley), 174

meditation, 106–7, 110, 112, 113–16,
117n8, 127–31
memory: cell theory of cerebral
physiology and, 50–51, 60, 62, 64;
collective, 194; computer and brain
compared, 230; disorders of, 204–5;
neuroscience of, 74, 112, 185, 189,
201, 226; philosophy of, 81–85; selfhood and, 90
Mesmer, Franz Anton, 31–33
mesmerism, 30, 34–37, 39–45
metaphor: brain as, 192; of the brain,
62; neural, 71–85; neuroplasticity as,
15, 207; transdisciplinarity and, 87
mind: affective and social neuroscience
and, 195; Alzheimer's disease and,
211; limitations of imaging technology and, 180–95; literature and,
216–31; primacy in the neuroscientific turn, 180; reading, 1, 36–37. See
also mind/brain relationship
mind/brain relationship, 10, 12, 15,
91, 144, 146, 199–202, 211. See also
dualism
mindfulness meditation, 124–25, 127–
28
mirror neurons, 72, 74–75, 77, 80–82
Mitchell, S. Weir, 30
Mol, Annemarie, 237

Nagel, Saskia, 11, 21n9, 27
narrative medicine, 95–99, 211–12
National Institute of Mental Health
(NIMH), 206
naturalistic fallacy, 143–44
Negri, Antonio, 217
neologisms, xi-xii, 153. See also neuro-
prefix
neural metaphor: described, 72; operation of, 74–76, 79
neurasthenia, 36–37
neuro- prefix, 8–13, 21n14, 65, 153,
202–3. See also neologisms
neuroaesthetics, 2, 8, 15, 182, 194
neuroculture, 21n14, 90, 92
neuro-disciplines, 2, 8, 20n5
neuroeconomics: critique of, 163–77,

182; defined, 169; discipline of,
20n2; 182, 238; language and, 14;
neologisms and, x, xi, 2, 8; normative, 170–74; positive, 169–70; value
judgments and, 171–77
neuroessentialism, 18, 19, 21, 146. See
also reductionism
neuroethics: bioethics and, x, 135–36,
143–44, 201, 203–5; clinical applications of, 136, 145–46; critiques
of, 10, 141–47, 200, 203–4, 205;
defined, 136; moral development
and, 139–40; moral reasoning and,
138–39; personhood and, 140–41;
pragmatic theory of, 135–48; social
implications of neuroscience and, 3
neurofiction: contribution to neuroscience, 99–101; critique of, 91–94;
defined, 90. See also neuroliterary
criticism; neuroliterature; neuronovel
neurohistory: critique of, 155, 236;
defined, 106–7; neologisms and, 2, 8
neuroimaging. See imaging technologies
Neuro-Journalism Mill (James S.
McDonnell Foundation), 192,
195n3. See also popular neuroscience
neurolaw, xi, 20, 182, 203, 239
neuroliterary criticism, 6–7, 90. See
also neurofiction; neuroliterature;
neuronovel
neuroliterature, 6, 21n11, 21n12.
See also neurofiction; neuronovel;
pedagogy
neurological determinism, 141–42. See
also reductionism
neurology: fiction and, 89–101; history
of, in nineteenth century United
States, 29–45; journals, 93, 95–97;
neuroplasticity and, 207; neuroscientific turn and, 158, 199–213;
professionalization of, 31, 40–45
neuronovel, 91–93. See also neurofiction; neuroliterary criticism;
neuroliterature
neuroplasticity, 15, 22n20, 22n22, 122–